A FAVORED PLACE

Frontispiece: Across the San Juan Basin, looking northeast, early in the rainy season. The landscape is located and shown from the air on figure 1.1. The palms belong to what remains of a grove on a pasture named El Yagual, after the palm commonly known as *yagua* (*Roystonea dunlapiana*). The striation suggested does not necessarily pertain to the remains of Prehispanic planting platforms.

A FAVORED PLACE
San Juan River Wetlands, Central Veracruz, A.D. 500 to the Present

by Alfred H. Siemens

 University of Texas Press, Austin

Requests for permission to reproduce material from this work
should be sent to Permissions, University of Texas Press,
P.O. Box 7819, Austin, TX 78713-7819.

⊗ The paper used in this publication meets the minimum
requirements of American National Standard for Information
Sciences—Permanence of Paper for Printed Library Materials,
ANSI Z39.48-1984.

Library of Congress Cataloging-in-Publication Data

Siemens, Alfred H.
    A favored place : San Juan River wetlands, central
Veracruz, A.D. 500 to the present / by Alfred H. Siemens. —
1st ed.
        p.   cm.
    Includes bibliographical references and index.
    ISBN 0-292-77727-2 (cl : alk. paper)
    1. Wetland agriculture—Mexico—San Juan River Region
(Veracruz)—History.   2. Traditional farming—Mexico—
San Juan River Region (Veracruz)—History.   3. San Juan
River Region (Veracruz, Mexico)—Antiquities.   I. Title.
S604.4.S54   1998
630′.972′62—dc21   97-43415

For Bill Denevan
scholar and friend

Contents

List of Tables   viii

List of Figures   ix

Portrait   xiii

Preface: An Amiable Collaboration   xv

1. Finding and Deciphering Patterned Ground
   in the Lowlands of Mesoamerica   1

2. Deducing the San Juan Basin in A.D. 500   45

3. Probing the Ethnohistorical Literature Surrounding
   the Encounter   101

4. Reformatting Sixteenth-Century Documents   110

5. Maximizing Some Late-Eighteenth-Century
   Observations   148

6. Appreciating a Naturalist's Rendition of Central
   Veracruz in the Nineteenth Century   160

7. Struggling with a Technocratic Pathology of the Basin
   in Mid-Twentieth Century   196

8. Summing Up the Yields   251

References   273

Index   291

List of Tables

2.1. Ceramic Counts from Remains of Platform in *El Yagual* 93

7.1. Land Tenure 220

7.2. Land Use in the La Antigua Irrigation District Early in the 1970s 228

7.3. Crops in the La Antigua Irrigation District, in Hectares 230

7.4. Approximate Calendar of Annual Crops 231

7.5. Excerpts from the Statistics of Participation in the Day of Demonstration 237

7.6. Animals in 1973 241

List of Figures

Frontispiece: Across the San Juan Basin, looking northeast,
early in the rainy season   **ii**

1.1. Views of the region:(a) Central Veracruz;
(b) the wetland pasture known as El Yagual   **2**

1.2. Locational map with the phases of the investigation   **9**

1.3. An example of the patterning to be found in
wetlands around the Candelaria River, Campeche   **13**

1.4. Remains of canals; probably a part of the waterborne
transportation infrastructure of Acalan   **17**

1.5. Remains of Prehispanic fisheries   **19**

1.6. Patterning in a *bajo* near Chetumal, Quintana Roo   **23**

1.7. A *bajo* in the Petén   **25**

1.8. A typical series of features along the Estero de
Tres Bocas, not far north of Nautla   **37**

1.9. Heuristic model of wetland margin   **38**

1.10. Variations in the topographic expression of the
elements shown in figure 1.9   **41**

1.11. Profile of an actual wetland margin near El Palmar
during the dry season   **42**

2.1. Aerial view of a part of the flooded San Juan Basin   **46**

2.2. Hills and lowlands west of Veracruz, with
preintervention streams   **47**

2.3. Distribution of Prehispanic patterning in its
topographic context   **49**

2.4. Physiographic overview of Central Veracruz   50

2.5. Geology of Central Veracruz   51

2.6. Sample of the vegetation over the patterning   54

2.7. Representative climatic graphs   55

2.8. Isohyets over Central Veracruz   56

2.9. Average annual fluctuation in the level of the
       La Antigua River at Cardel   58

2.10. Hydrography before the intervention by the
       Secretaría de Recursos Hidráulicos   60

2.11. Floodwaters in the San Juan Basin in July
       of 1993 superimposed on the distribution
       of patterning   63

2.12. Seasonality in and around a tropical lowland
       floodplain   66

2.13. Locational map for figures 2.14–2.21   71

2.14. Air oblique photograph of the Tulipan complex   73

2.15. Canal midlines for the Tulipan complex   74

2.16. Air oblique photograph of the Carmelita complex   75

2.17. Canal midlines for the Carmelita complex   76

2.18. Air oblique photograph of a part of the Nevería
       complex   78

2.19. Canal midlines for the Nevería complex   79

2.20. Topographic map of the Caño Prieto settlement site   80

2.21. Air oblique photograph of stone lines on the
       piedmont near Rinconada   81

2.22. Map of stone lines on the piedmont   82

2.23. Canal midline orientations in the wetlands
       of Central Veracruz   89

3.1. Depopulation in the Gulf Lowlands from 1520
       to 1570   108

4.1. Profile of vegetation around Old Veracruz (La Antigua)   124

4.2. Profile of vegetation in the western San Juan Basin   127

4.3. Profile of vegetation on one side of a wide *barranca,*
in the hill land west of the Basin   129

4.4. An early colonial land-litigation map   130

4.5. The map evoked by the early colonial *Declaración,*
Relación, and related materials   132

5.1. A map of Central Veracruz in 1784 by Miguel
del Corral   153

5.2. Central Veracruz by Alexander von Humboldt   154

5.3. A measured profile from Mexico City to Veracruz   157

6.1. A schematic profile of Central Mexico   170

6.2. The banded mainland of Central Veracruz as seen
from the sea   171

6.3. Sartorius's hypothesis on sedimentation along
the Central Veracruzan coast   172

6.4. East-west and north-south profiles of live and fixed
dunes along the coast of Central Veracruz   172

6.5. Engraving of Moritz Rugendas's drawing of vegetation
on an "island" within the San Juan Basin   176

6.6. Profile of the typical relationship between the
eastern extremes of the piedmont and the beginning
of the *barrancas*   178

6.7. Profile of piedmont surface littered with Prehispanic
remains and a *barranca*   178

6.8. Profile of a narrow barranca   180

6.9. Engraving of Rugendas's drawing of Indians against
their house in Xalcomulco   183

7.1. Map of the hydraulic intervention of the Secretaría de
Recursos Hidráulicos in the San Juan Basin   198

7.2. President Miguel Alemán   206

7.3. An illustration of a fundamental Mexican
statistical problem   216

7.4. Settlements in the San Juan Basin in 1970   217

7.5. A schematic profile of topography and geology,
with soil series referred to in the text   222

7.6. (*a*) Roundup of cattle on a medium-sized lowland
Veracruzan ranch in 1961, and (*b*) the owner of the
ranch posing with his recently acquired zebu bull   244

8.1. Prehispanic patterning superimposed on land tenure
in 1970   258

Portrait

Alfred H. Siemens, professor of geography at the University of British Columbia. Author of *The Americas* (1977), *Tierra Configurada* (1989), *Between the Summit and the Sea* (1990). Current research: reassessment of Prehispanic land and water management along the Candelaria River of Campeche, Mexico, and analysis of the impact of, and resistance to, contemporary agricultural globalization. Collector of Latin American folkloric dancers' masks.

Preface

An Amiable Collaboration

The long involvement with this one place grew out of shared enthusiasms as much as personal absorption. Mario Navarrete Hernández, an archaeologist and Veracruzano, entered into the study with me. La Antigua is his home town; he helped me appreciate it and taught me a good deal about the prehistory and the history of the lowlands to the south of it. We poured water over each other's heads at the end of a day's excavation. When he was made director of the Museum of Anthropology in Xalapa and took his place on the national council overseeing archaeological projects, he was able to help our joint investigation of Prehispanic wetland agriculture through various bureaucratic thickets.

Groups of men from the villages of El Pando and Loma Iguana helped us in our fieldwork, commenting on their surroundings or the events of the day and recounting sometimes how things had been. The *gringos,* in turn, often amused them. I have thanked them personally long since, of course, but acknowledge their assistance here as a reminder of some of the most enjoyable aspects of fieldwork.

This project could not have been carried out if Ing. Carlos Soza Lagunes and Don Francisco González had not graciously allowed us repeatedly to perforate their pastures.

Arturo Gómez-Pompa, when he directed the Instituto Nacional de Investigación sobre Recursos Bióticos, showed warm personal interest and provided most agreeable contexts for the early discussion of the possibilities for the investigation of Prehispanic wetland agriculture in Central Veracruz. He activated a key contact in a federal ministry for the release of materials on the San Juan Basin. Various members of the institute participated in the investigation; Manuel G. Zolá Báez was often in the field with us and contributed valuable botanical and microtopographic materials.

Alastair Robertson, who should have been a co-author, was drawn even further than I into the San Juan Basin's detail. I value his work with air photographs, the large-scale maps of the agricultural remains that he prepared, and his determined search through repositories of governmental materials, but most of all his close environmental observations. I see him still, on his motorcycle, trailing exhaust, with some new measuring device of his own fabrication tied to the rear carrier. He shared the passion, took endless pains, and on occasion held up a mirror.

Julie Stein, archaeologist at the University of Washington in Seattle, taught me about coring. She did not need to come into the field for this; she did it in periodic "tutorials" in her office and meanwhile published basic essays on the subject. Dolores Piperno of the Smithsonian Tropical Research Institute helped us a great deal by interpreting the phytoliths in our samples.

Richard Hebda, paleoecologist at the Royal Museum of British Columbia in Victoria, and adjunct professor at the University of Victoria, made time repeatedly to come into the field. He led us all in the interpretation of the natural context of our Prehispanic materials and the planning of the processing of the various diagnostic remains, but that is too fibrous a way of putting it. He was, he is, always positive, energetic, and often entertainingly expansive.

I thank Richard's wife Elaine for her forbearance, and my wife Alice, not only for her forbearance but her active support at many junctures.

A variety of grants, from the Social Science and Humanities Research Council of Canada, the National Geographic Society, and the Humanities and Social Sciences Committee at the University of British Columbia, helped to sustain us.

A number of Mexican discussants lent me their perspectives: Juan Pablo Martínez Davila, instructor at the Colegio de Postgraduados, Campus Veracruz, helped me to understand the mid-twentieth-century reports of the engineers of the Secretaría de Recursos Hidráulicos (SRH) regarding the San Juan Basin. I was fortunate to spend time with various of his colleagues too, particularly Carlos Olguín Palacios and Octavio Ruiz Rosado. The still sprightly Ing. Adolfo Orive Alba, minister of SRH during the presidency of Miguel Alemán, when the major hydraulic intervention in the Basin was undertaken, shared some fine background political information and most interesting historical photographs. Ana Lid del Angel Pérez, investigator at the Instituto Nacional de Investigaciones Forestales y Agropecuarias, Campo Cotaxtla, Veracruz, was free with the results of her inquiry into animal husbandry on the eastern side of the Basin.

Andrew Sluyter, geographer at the University of Texas at Austin and then at Pennsylvania State, has long agreed with me that Central Veracruz is the center of the world. He obtained a fine long core out of the central lake of the San Juan Basin and thus provided paleoecological context and finer meshed elaboration of our introductory efforts. He mastered not only the materials of prehistory but also those of history; he facilitated many specific aspects of my own historical research. His doctoral dissertation on the ecological restructuring of Central Veracruz after the Encounter is state of the art—and the polemic.

Alba González Jácome, anthropologist and historian at the Universidad Iberoamericana in Mexico City, facilitated our work with contacts, advice, and criticism—all from a solid base of scholarly integrity and good sense, and often enough with pointed repartee.

Tonatiuh Romero, graduate student at the Universidad Iberoamericana (IBERO) and archival adept, guided me through the Archivo General de la Nación in the Palacio Lecumberri, the old prison straight out of the eighteenth-century plans of Jeremy Bentham; we would begin in the central rotunda and go down one or another of the spokes of the great wheel into the relevant sections of materials accommodated in former cells.

I was able to ventilate the techniques and concepts of the whole long inquiry into Prehispanic wetland agriculture in various graduate seminars at IBERO. I came to appreciate the students very much; they were kind enough to repeatedly declare themselves interested, even enthusiastic.

There were other forums, notably the meetings of the Scientific Committee on Problems of the Environment (SCOPE) in Seville, in January of 1992, focused on the environmental impact of the Encounter. Such an inquiry is strongly indicated for Central Veracruz and the Mexican Gulf Coast, of course, but the results cannot be as sweeping or dramatic as they appear to be in some other areas of Latin America. In the preparation of an edited volume in honor of my good friend Phil Wagner, I was able to sketch out my idea for the treatment of the Basin (Siemens 1992). Odile Hoffmann invited me to a lively symposium on the social organization and representation of space at the Centro de Investigaciones y Estudios Superiores en Antropología Social, Región Golfo, at Xalapa, Veracruz, in September of 1994; I was able to test out some of Foucault's ideas in Central Veracruz.

David Skerritt Gardner, a historian at the Universidad Veracruzana and intimately familiar with matters agrarian in Central Veracruz, invited me to a seminar on regional processes, but also repeatedly met me for coffee at La Parroquia in Xalapa. He tempered and supplemented many of my historical observations, indeed he took the trouble to read an early version of this analysis and made very good suggestions. I am grateful also to William Doolittle, geographer at the University of Texas at Austin, for his encouragement of this project and his reading of the manuscript. Other encouragement came out of Austin: Shannon Davies has been a gracious, facilitating editor. She managed to sound surprised at one point that the final version would be ready so soon. Various of her colleagues helped this very imprecise author with the minutia of copyediting.

From the beginning of this project I followed a chimera: computerized wizardry would somehow allow me to do marvelous things, like making vertical photographs out of obliques and converting the San Juan Basin into dramatic topography that could be tilted, rotated, and inscribed with any number of distributions. I was defeated on both counts, always just ahead or just behind the necessary technology, dogged by incompatible map bases or software versions. I came to appreciate a fine Mexican expression: "sí, pero no"—yes it can be done, but unfortunately not. Catherine Griffith helped me get what was feasible and is the main architect of the maps and diagrams; Daniel Millette assisted at several critical cartographic junctures, as did Elaine Cho, Solomon Wong, and Doug Brown. Maija Heimo and Klaudia Zapala amiably helped to tame the text.

1. Finding and Deciphering Patterned Ground in the Lowlands of Mesoamerica

An open window, or better yet an open door in a light plane: the ultimate panopticon.

This will be a close look at some rather flat terrain just inland and northwest of the port of Veracruz. Most of each year it is drained northward by the San Juan River, through a string of wetlands into the west-east flowing La Antigua River just before that larger stream reaches the sea. In the wettest times of the year the water is likely to flow in the other direction. I have labeled this landscape the San Juan Basin or just the Basin. It is not easily recognized on a map or on the ground, but it does cohere and has proved absorbing for the Prehispanic remains in its wetlands, and for what overlies them (figure 1.1).

It would have been gratifying to have predicted that the wetlands in the lowlands of Central Veracruz would be as intriguing as other wetlands in the Mesoamerican lowlands had already proved to be and then to have planned an exploration. Instead, the involvement was fortuitous. A commercial flight on December 13, 1976, from Mexico City to Minatitlan was not allowed to land because of poor visibility and diverted to Veracruz. On approach to that airport the cloud cover gave way and directly underneath us lay a strikingly patterned wetland—patterned in its vegetation, with rectilinear forms among the irregular and the circular. Shortly thereafter more systematic aerial reconnaissance showed that there were various major complexes of such patterning in the region, including a number in the San Juan Basin.

Various colleagues and I were eventually able to confirm that this represented a Prehispanic agricultural system and that it functioned here around 500 years after the time of Christ (Siemens et al. 1988). Similar remains had already been found in the Maya region; the inquiry had now been projected westward. It has since been possible to give it new facets and to combat misunderstandings. It continues to yield, even though, or perhaps precisely because, certain aspects of it have become clichés.

Fig. 1.1. (a) Central Veracruz; (b) view over the wetland pasture known as
El Yagual, site of investigation reported in Siemens et al. 1988. This is a
subset of the larger Nevería complex, shown in figures 2.18 and 2.19.

A number of authors have ably assessed the current state of play and elaborated the conceptualization. Fedick, for example, has traced advances in the study of indigenous agriculture in the Americas generally (1995). Scarborough and Isaac have brought together reexaminations of the economic aspects of water management in the Prehispanic New World (1993). Sluyter has reviewed the investigation of intensive wetland agriculture in Mesoamerica in detail and sees a series of important tasks ahead (1994: 575–576). Stanish reconsiders the hydraulic hypothesis, that venerable explanation of agricultural intensification, and affirms that "It is not simple hyperbole . . . that raised-field agriculture is among the most important intensification strategies utilized by Prehispanic farmers in the Americas" (Stanish 1994: 312).

However, it is not just the patterning that has stimulated this analysis. The landscape presents some strong juxtapositions, epitomized by superimposed canals. Modern drainage canalization, with its fairly clear rationale, overlies an earlier canalization not nearly so clear, visually or functionally. What intervened? What relationship does the ancient system have to current land use?

Historical references to this and other wetland regions often deprecate them strongly and yet they were once sustaining areas. How did they change, or is it a matter of representation?

Crossing the cracked earth of the wetlands of the Basin in the dry season one sometimes has the feeling that much had taken place within earshot. Prehispanic hegemonies, the Conquest, colonial trade, the nineteenth-century crossings of the lowlands by foreign observers, all this swirled around to one side or the other. In the twentieth century hydraulic engineers attempted to rearrange the hydrology of the Basin itself and facilitate modernized agriculture. Some historical currents thus pass by, others enter; there are continuities as well as discontinuities. The Basin thus fairly invites expansive diachronic consideration in addition to the attention we have given it for the time of the actual use of the planting platforms and canals. Varying interactions of people and a suite of biomes, or of introduced animals and a set of topographic, hydrological, and vegetational conditions, lend vitality; a biographical metaphor is thus not entirely outrageous.

Reducing the objective of this study very deliberately: it is to consider the context, nature, and fate of production systems operative within and just around the wetlands of Central Veracruz, from Prehispanic times to the present.

The Nature of the Analysis

The verbs in the chapter titles signal something of the work required. Air reconnaissance opened the search in the lowlands and successively enlarged it, affecting the gist of the analysis fundamentally and providing essential visual material.

The view is at an angle somewhere between that of the observer on the ground, usually surrounded in tropical lowlands by obscuring vegetation, and that of the vertical air photo camera mounted on the floor of a fuselage. The perpendicular view, a fundamental novelty for a human observer, yields a landscape in symbols. The characteristic forms, tones, and juxtapositions of the area under consideration must be learned before photos from that perspective can be interpreted. Taken in closely regulated series, they come alive under a stereoscope and allow measurement.

The oblique view, on the other hand, is more familiar, more realistic; it is only an exaggeration of the sort of view one has from any eminence. One cannot measure sizes or directions on an oblique photograph; scale varies continuously from the bottom to the top and directions converge. However, it accommodates the easy shifts in the direction of view and change of altitude that are possible during light-plane reconnaissance. What is seen is easily readable but remarkable.

Colonial Spanish American buildings often had a *mirador*, a lookout for recreation or supervision (Siemens 1994). This is easily linked to Jeremy Bentham's or Foucault's ideas on the panopticon, the place of normative, dominating observation in all directions (1962; 1979). On a clear day at several thousand feet above the lowlands one often feels one has that sort of privileged view.

A metaphor often obtrudes on the examination of a landscape by these means. Flat alluvial plains such as those of the San Juan Basin, inscribed by those who dug the early canals and heaped up the planting platforms, as well as those who added their marks later, can easily seem a text. The alphabet of the Prehispanic agricultural and transportational inscription is wetland vegetation— the differences in tones or hues consequent on microenvironmental variation. This is often superscribed by the geometry of recent agriculture and ranching. The concept becomes almost tangible when one has a graphic representation in hand. However, it is also malleable. My own analysis here of such a landscape text is a literal text. It must rely on those of previous authors. We are thus very much between texts. The interplay and the yields have been elaborated well by Barnes and Duncan, who, as many of us now, use the term in an expanded sense:

> One that includes other cultural productions such as paintings, maps and landscapes, as well as social, economic and political institutions . . . This expanded notion of texts originates from a broadly post-modern view, one that sees them as constitutive of reality rather than mimicking it—in other words, as a cultural practise of signification rather than as referential duplications . . . such practises of signification are intertextual in that they embody other cultural texts and, as a consequence, are communicative and productive of meaning. Such meaning, however, is by no means fixed; rather, it is culturally and historically, and sometimes even individually and momentarily, variable. (1992: 5–6)

The analysis, and signification, has usually proceeded from air to ground reconnaissance, including some free-form ethnography. One could call this the quintessential introductory geographical ponder. It can hardly be systematically described or given decent structure in a formal research report, but it is invariably instructive and generates hypotheses.

Scientific examinations of the stratigraphy under the patterning and of its environmental context have followed sooner or later. This has always required much consultation and often led to collaboration; authorship must therefore often be referred to in the plural.

The examination of the relevant primary documentary materials, particularly those to be found in the various treasuries left by the Spanish colonial administration, can almost seduce one away from field inquiry. Here are texts indeed. Then there are the repositories of twentieth-century official papers, particularly those of the Secretaría de Recursos Hidráulicos (SRH), which are much less seductive. All these materials have had to be reformatted—word-processing is both the analogue and the means—to make them useful.

Assaying the nineteenth century leads into the publications of foreign observers. Those made by naturalists are already quite congenial in their philosophical terms of reference. Central Veracruz is fortunate to have such an observer, Carl Sartorius, as we will see in chapter 6. The corpus as a whole, the accounts of the diplomats, the speculators, the soldiers who came to enforce one hegemony or another, all that sketches us a region in a particular period but is also laced with strong preconceptions, which invite deconstruction (Siemens 1990).

Our own observations, foreign too, after all, have been informed by lively discussions with resident scholars and their various publications. In recent years there have been thoughtful new critical analyses of the region and its history, offering useful concepts and perspectives (e.g., Skerritt 1989; Hoffmann and Velázquez 1994).

Yields

The analysis has yielded in two main ways. Firstly, even though the inquiry is variegated, there are some realizations that can be spliced in a fairly linear fashion. It is possible to come to some conclusions about the nature, function, and chronology of the wetland agricultural system, the land use in its immediate surroundings as well as the natural environment of it all. The altitudinal rack on which the adaptations of the various epochs are stretched can be detailed. Most importantly, perhaps, the tendencies in the use of land can be elaborated in new ways.

Secondly, the inquiry recurves in the representations that it must sample. One is led, for instance, to reflect on the nature of scientific inquiry in the various epochs: the Hippocratic context of the *Relaciones Geográficas*, the more comfortable expositions of the nineteenth-century naturalists, open to the emergent ideas about process, then the categorical and rather arid science of those who wrote the reports of the SRH, and finally the scientific aspirations of our own inquiry, avid for ecology.

One contrast obtrudes: historic assessments of tropical wetlands have long been highly deprecatory, and understandably so. Cortés, struggling eastward with his huge retinue through the Gulf Lowlands toward Honduras, expressed it well at one point when they came on a great swamp two crossbow shots wide: "The most terrible thing that man ever saw" (cited in Scholes and Roys 1968: 105). Not only daunting logistical deterrents, wetlands were also considered sources of mortal illness. Humboldt himself has some of the best passages on this in his discussion of the lowlands of Veracruz. The only good wetland, he and others implied, was a drained one. Over against that is the evidence for Prehispanic intensive agriculture on the margins of many wetlands, implying a positive evaluation, a bending of effort and ingenuity to use these environments, to ride with their seasonal rhythm.

The contrast has echoes in recent decades. There can be no doubt that wetlands are key to contemporary cattle ranching; cattlemen strive to drain them up to a point, but not completely. They are advantageously exploited here and there for seasonal agriculture as well. But for decades, governmental planners have talked and practiced drainage without much discrimination, ignoring the inherent potential of the wetland and remaining fairly insensitive to traditional use.

Any inquiry into wetlands now is overshadowed by strong current perceptions. The aesthetics of wetlands are widely appreciated and are well exploited in television documentaries. Artifacts and bodies dropped into them remain well preserved—the archaeology of wetlands has had spectacular results. The prospects of further contamination, of drainage, and of the building over of wetlands large and small in many parts of the world are highly worrisome. Knowledge of the incredible productivity of this biome and its important hydrological function is gradually diffusing.

The Perspective Taken

From the beginning of our consideration of the remains of ancient wetland use it has seemed logical and conceptually productive to hypothesize advantage; the wetlands held and hold a very positive sustaining potential. This book explores the ways in which this favor has been realized.

A German geographer, Franz Tichy, observed some years ago that for the Prehispanic occupants of Central Mexico the margins of the shallow upland lakes were particularly favored terrain. Residual moisture there could sustain agriculture during the dry season (Tichy 1979: 343–344). It seems justifiable to appropriate that fortunate designation for wetland margins in the lowlands as well.

A Hapless Addendum

Those of us working on Prehispanic wetland agriculture have often expressed the hope in grant applications and elsewhere that our investigations would have some practical significance, that we would be able to make some actual recommendations for the enhancement of production in the regions in which we worked and that some of the ancient favor could be realized again.

We see the lapse of traditional agricultural systems in the lowlands, the failed developmental projects, the environmental contamination, the marginalization of many rural people—pauperization is often a more accurate term—and the out-migration, in short, all the various unfortunate aspects of modernization and globalization. Considering the remains in the wetlands encourages reflection on the juggernaut and possible alternatives (Siemens 1996).

The necessity of the protection and non-destructive use of wetlands is obvious. It is advisable to work with wetlands more or less as they are rather than massively rearranging them. The various ancient productive activities that may be deduced for the wetlands and their immediate surroundings reinforce the importance of agro-ecological diversity. This, in turn, may contribute to the discussion of the troublesome issue of sustainability. No production system lasts indefinitely, but one that appears to have lasted for centuries must have some lessons to teach in a context in which initiatives frequently wax and wane in a matter of decades.

The model of the production system that we propose toward the end of this chapter could provide a basis for experimentation. Experiments in wetland agricultural use have been undertaken, of course (e.g., Gliessman et al. 1981), but not with the specific parameters of this model. Colleagues working in the environs of Lake Titicaca have been very successful in this respect; they have been able to facilitate the reactivation of raised fields (Erickson 1985; Kolata et al. 1989). In the Mexican Gulf Lowlands there are great differences between the socioeconomic context of today and what we can reconstruct for the time of the use of the planting platforms and canals. Furthermore, most of the patterned wetlands of today are also not accessible to those who would like to till them. Relevance is thus not immediately apparent.

The results of our work cannot yet, or may not ever be, as specific as the questions asked of it by practical agriculturalists. We have few immediate and explicit suggestions to make, things that could be tried in the next cropping cycle. The whole phenomenon of ancient wetland agriculture is likely to be seen by the agriculturalist and the rancher as a curiosity. Its relevance lies at a general level, challenging policy-makers to reflection and encouraging resistance to the prevalent trends in food production.

Phases of the Larger Search within the Lowlands

Finding the patterning in Veracruz in 1976 was a continuation of a protracted search (figure 1.2). The rectilinear marks in amorphous contexts had been found in 1968 in some of the wetlands of the Maya lowlands (Siemens and Puleston 1972; Siemens 1989c: 55–83). In repeated forays their more specific geography became apparent. Some were found on the edges or across backswamps, others

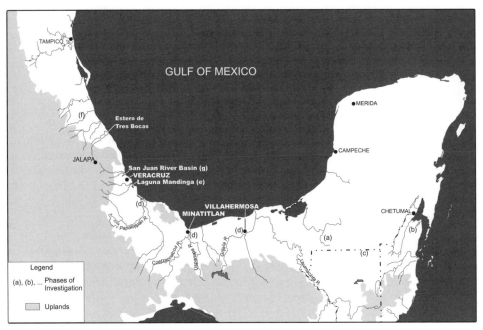

Fig. 1.2. Locational map with the phases of the investigation of wetland patterning in Mesoamerica.

on the gentle landward slopes of levees, still others across silted meander bends, and many, particularly in the San Juan Basin, on the gently sloping, partly colluvial and partly alluvial slopes between wetlands and terra firma. The variations are dealt with more specifically and graphically near the end of this chapter. The common characteristics of these micro-environments were that each provided a store of moisture that could be tapped in the dry season and the flooding across them in the wet season was usually benign. These conditions most often obtained on the margins of wetlands; this *ecotone,* or transition between two ecological communities, therefore deserved particular attention.

Evidence of agriculture in the wetlands was a substantiation of suspicions raised years ago by Angel Palerm and Eric Wolf:

> The abundance of swamps and lakes in the lowlands and the apparent Mayan predilection for such surroundings, suggests the possibility of a system of cultivation of swamplands, using techniques resembling those of the chinampas of the highland . . . Classic Maya agriculture may have been a very com-

plex and varied system, which emploved a variety of techniques, each adequate in terms of certain specific local conditions. We can find no valid reasons for the assumption that
slash-and-burn agriculture was the only or even the major
technique employed. (Palerm and Wolf 1957: 28–29)

This turned out to be a shrewd formulation. It set into relief the
increasingly uncomfortable basic hypothesis regarding Maya subsistence, that shifting cultivation had been its mainstay. There had
to have been something more. Terracing had long been indicated
for the Belize River valley; soon its vestiges were to be found in
other locations in the peninsula. The patterning noted in the wetlands along the Candelaria indicated a system that was indeed
chinampa-like but not yet fully *chinampas*. This system involves
a cutting of canals into the margins of lakes or floodplains, building up planting platforms, introducing fresh water on the upslope
margins and somehow preventing flooding from downslope, thus
maintaining water levels and allowing intensive agriculture, with
an array of intriguing specifics, all year round.

The *raised fields* indicated by the patterning in many lowland
wetlands are physically a network of canals and intervening planting platforms too, but they were generally subject to seasonal inundation—a significant difference in the degree of hydrological
control, implying many differences in actual practice and perhaps
a developmental process, from the less to the more intensive. It
may be useful occasionally to use something like *proto-chinampa*
as a synonym for *raised fields*, or simply *planting platforms and
canals*. In any case, the remains were taken as evidence for intensive agriculture, which seemed to allow a more satisfactory interpretation of Prehispanic subsistence than had been available up to
that time.

Basically, the statement by these thoughtful observers of Middle
American agriculture evoked the dynamics of differentiated and
interdigitated agricultural activities: one crop underway in one
field at a particular time, a starting bed of other plants ready for the
move into a neighboring plot when the time is right, crops following a descending water level perhaps, harvests just managed in the
lowest areas, with canoes if necessary, as the water rises again, all
these activities intensified if more is needed and curtailed if not.
Such congeries can hardly fail to appear as highly productive; the
landscapes in which they occur seem finely tuned. In this lies the
favor of the wetlands.

Southwestern Campeche (figure 1.2, no. 1)

A review of the oblique imagery obtained during a critical flight over the environs of the Candelaria River in 1968 shows a perceptual shift that facilitated the discovery. The flight was intended to update information on recent new settlement, and a series of images do show the communities and lands of this venture. Then a film had to be replaced—always a potential break in the gist of observation. After the new film was in, the view had changed. The first frame is dominated by straight lines across the floodplain, apparently unrelated to current agriculture or ranching. Attention had been redirected; soon rectilinear webs appear in the photographs as well. Not long before that flight, James Parsons, a geographer at Berkeley, had showed me results of wetland reconnaissance that he and others, particularly William Denevan, had already carried out in Andean South America, and had remarked that surely something like the very impressive Prehispanic patterning they had found would surface somewhere along the Gulf of Mexico. Here it was; Palerm and Wolf's prediction was substantiated.

It was possible to carry out fairly extensive ground reconnaissance and a few test excavations, but further work was obviously needed. Funding was available but a permit was denied. We were given to understand that it would not be appropriate for foreigners to work in the environs of Itzamkanac, the site along the Candelaria River, where the bones of Cuauhtemoc were believed to be buried. Behind this, no doubt, were questions of territoriality; we were seen as intruders. In short order we proposed an investigation to the authorities in Belize, where we had already done air reconnaissance and found extensive patterning. We were given permission and changed our venue.

Although the vestiges in the basin of the Candelaria River could not be investigated further on the ground, the available vertical and oblique photography was subjected to a rather determined scrutiny (Siemens 1989: 31–83). Also, the subsequent work on the same phenomenon in other locations reflected back on the early results along the Candelaria. Some early naiveties could thus be sluffed off; other early ideas proved well founded and could be elaborated; misconceptions could be more firmly countered. In 1994 the professional territoriality that had put the basin off limits could be evaded. Further fieldwork has since been carried out and, although the results are far from completely analyzed, some new findings can be adumbrated (Siemens 1995).

The basin of the Candelaria River above the rapids and ancient dams just to the west of the town of Candelaria coincides roughly with a Prehispanic culture region: Acalan. The name is derived from *acalli,* which means "place of the canoes." It was the homeland of the Putun Maya (a subgroup of the Chontal Maya), which emerges for us from the excellent analysis of early Spanish documents by Scholes and Roys (1968). Air photography allowed us to identify some of its seventy-six communities in the Candelaria River basin. Many mound groups had become apparent with the recent clearing of land for pasture and crops. One cultural landscape was emerging from underneath another.

A much larger group of mounds was obvious, those of Itzamkanac, now often referred to as El Tigre. It is the focus of a longterm investigation by archaeologist Ernesto Vargas Pacheco and his team (Vargas 1995), to whom we also owe a synthesis of the Prehispanic history of the region (Vargas 1994). A good deal may be expected of this investigation; it is already apparent that the chronology of the site goes back to the Pre-Classic, which gives ample room for placement of the chronology of the wetland agricultural system. The bank of the Candelaria River near the site shows remains of the port that is mentioned in the early Spanish documents; impressive ceramics have been retrieved out of the channel itself (Vargas, personal communication, 1996).

There is no mention of wetland agriculture in the documents, which does not necessarily mean that it was not practiced. The region much impressed the first Spanish visitors by its productivity; they were able to obtain provisions here for their considerable company on the incredible march toward the Gulf of Honduras. It has been pointed out that at the time of the arrival of the Spaniards, the Chontal Maya were already in decline (Vargas 1994: 61). This probably has a good deal of significance for the interpretation of the remains of wetland agriculture in this region in that it allows the postulation of agricultural disintensification before Conquest, here as in other lowland regions.

Agricultural Remains.   The rectilinear forms apparent among the rounded vegetation of this floodplain represent a latticework of short canals and the planting platforms between them, the *raised fields* (figure 1.3). Once their stratigraphy had been sampled here and elsewhere, they substantiated the intimation of intensive agriculture among the Maya and helped to clarify the troublesome question of their subsistence.

Fig. 1.3. An example of the patterning to be found in wetlands around the Candelaria River, Campeche. The hill lands around the wetland have the knobby topography characteristic of *karst*, i.e., terrain underlain by calcareous rock and subject to solution. The photograph was taken in 1994; much of the hill land had been cleared for pasture. The boundary of hill land and wetland is banded with forest growth, a common feature in the wetlands studied (cf. figure 2.6).

These vestiges proved to have typical relationships to topography and hydrology. They were found mostly on the margins of the lowest areas within the floodplain, the lakes or backswamps, the areas subject to inundation. Biologically this is an ecotone; our informants in various locations have often called it an *orilla,* a shore. It will be dealt with further near the end of this chapter.

Morphological and stratigraphic evidence in several locations has indicated that wetland agriculture with canalization and the heaping up of planting platforms—as, indeed, was the agriculture on unmodified wetland margins that was carried on instead of it or that preceded it in many places—was practiced subject to seasonal flooding, a basic and important point.

We were on the lookout from the beginning for evidence of hydrological control. There were some indications of control features, such as the traces of dams across *arroyos,* the seasonal streams coming out of field complexes, as though attempts had been made to retain water in the canals as long as possible during the dry season. But there was little evidence from our early forays of the extensive diking and freshwater feed-in from adjoining hill land that would have been necessary to maintain constant water levels for full-fledged chinampa agriculture. We still had much to learn from the local informants about springs and had not paid sufficient attention to structures across some wetlands that were referred to locally and in the literature as *sacbes,* i.e., ancient causeways. A new air reconnaissance in 1994 showed clear lineation behind several of the causeways, which the aquatic vegetation had happened to obscure on previous passes. Coring showed a clear agricultural horizon: maize pollen at something like a metre below the surface on the remains of the planting platforms.

Putting all this more recent evidence together, we were able to identify a series of slightly stepped wetlands just west of the main site, used for agriculture, enclosed by a succession of barriers, fed from upslope by various springs, and draining into the river. Hydrology had been controlled in wetlands right next to the main settlement and year-round cultivation would have been feasible. Thus, after all, there had been some full-fledged chinampas at this site, an exception that proved the rule. This added to the array of productive possibilities that could be imputed here and allowed the postulation of a further stage in the process of intensification.

On return to the field in the spring of 1996 we were made aware of hydrological control on another scale. It is common knowledge that the Candelaria River is impeded by rapids some distance downstream from the town. A local *aficionado* of the river and an-

tiquities, Angel Soler, made us aware of a series of artificial bar-
rages before the natural barrier, which he had sounded and mapped
meticulously. To one degree or another, then, the hydrology of the
whole basin upstream, the landscape of Acalan, as a matter of fact,
was being controlled. The implications and the chronology of this
control are now under study.

From the beginning of the investigation of the vestiges in the
Candelaria basin, we recognized that the webbed patterning in its
floodplain covered only a very limited total area when compared to
the massive examples found in South America or indeed elsewhere
in lowland Mesoamerica. They might have to be considered as a
curiosity that had no great significance for the interpretation of the
subsistence of the ancient Maya. On the other hand, it could be argued that within a given Prehispanic landscape the complexes of
planting platforms in the wetlands, although of very limited extent,
would have been highly productive and diagnostic of fundamental
developments, in particular agricultural intensification. The latter
has become the generally accepted interpretation.

Although the wetland fields have proved more than a curiosity,
they are still less than the single explanatory factor that popular
articles and television documentaries have made them out to be.
Also, the common label "intensive," applied without qualification, has oversimplified the academic discussion of wetland agriculture. "Raised field" has become a cliché. The discovery of Prehispanic agricultural vestiges should not be seen as providing a new
paradigm of subsistence, fully formed, so much as stimulating a
new sensitivity to the productivity of interdigitated cropping, with
its periodicity and responses to secular changes within the environment, its intensification and disintensification before Conquest.
And, it must be noted—one feels the point should be repeated a
hundred times in strategic places—that wetland agriculture in the
Mesoamerican lowlands was not solely or even primarily a Mayan
achievement. It extends well westward out of the Maya realm.

The Maya have often been aggrandized and taken as surrogates
for lowland culture and adaptation. The ingenuity of their subsistence system, and that of other peoples living in the lowlands toward the west, lay less in a vast hydraulic solution and more in an
orchestration of productive activities in neighboring microenvironments—and in the resulting capacity to adapt to seasonal and
secular climatic variations, particularly drought.

Other Canals. The Candelaria floodplain is inscribed not only
by rectilinear webs but by many longer straight lines (figure 1.4).

They represent canals; we were able to test this by trenching in several locations. Their interpretation benefits by the application of some very simple concepts of locational analysis, applied with maps and air photos in hand. This is "evidence" as "material" as a discontinuity on an excavation wall, a soil sample, or a potsherd, for that matter.

The canals are not usually related to present agriculture or ranching, but they are not necessarily related to the ancient agriculture either. They cannot be taken as direct evidence for "raised fields," which has happened more than once. Certain ones of them, by virtue of their directions, their termini, particularly their relationship to one or another of the many smaller Prehispanic settlement sites in the basin, make sense as an ancient waterborne transportation infrastructure. Similar lines, in smaller numbers, appear across the floodplains of Northern Belize; they are rare in the patterned wetlands of Central and Northern Veracruz.

Some of the lines that make up this infrastructure connect river banks with terra firma across an expanse of floodplain. These have their analogues in canals that we have seen recent settlers in the Candelaria basin cut and maintain in order to be able to get at land that can be used for shifting cultivation or to provide access from a small community on the edge of terra firma to the river. In places these have been cut through bundles of earlier canal remains, of which the modern builders were unaware. Prehispanic locational relationships are likely to have been similar in these respects.

There are lines that roughly parallel the river for long distances. They seem to have had a strategic significance. From recent work on Mayan glyphs has emerged a "lively story of politics and warfare" (Appenzeller 1994: 733). This probably applied for Acalan as it did for other Maya regions. Invaders coming along the main stream could be surrounded by defenders coming along canals concealed among the hydrophytes.

Some of the access canals we were able to trace focused in on settlements located and discussed in the early colonial accounts. As it happens, the recently reported remains of a port on the edge of Itzamkanac are about opposite to a roughly focused cluster of canals that we noticed on the northern side of the river when we first began the airborne reconnaissance and air photo analysis in this region.

There is certainly the possibility that some of the lines represent canals cut during colonial and post-colonial exploitation of dye wood in these parts (Vadillo 1994: 68–69). The interpretation of these various types of canals is thus no straightforward matter.

Fig. 1.4. Remains of canals; probably a part of the waterborne transportation infrastructure of Acalan; some of the lines focus on a group of low mounds off to the left—one of the many small communities of Acalan. In the background is one of the colonies established along the river in 1965. Figure 1.3 also shows what appear to be trans-wetland transportation canals.

There are also bundles of long, more or less straight lines that are not related to complexes of planting platforms and do not make sense as transportation routes, but do, instead, make a great deal of sense as Prehispanic fisheries (figure 1.5). They will be referred to again in chapter 2 in connection with the phenomenon of benign flooding on tropical floodplains.

It has been difficult to make the distinctions among the canals along the Candelaria stick. One prominent Mexican archaeologist, for instance—who was recently still discounting aerial evidence on the interpretation of Prehispanic use of wetlands in general and had not followed our distinctions in the chronological attributions made in the interpretation of the canals of the Candelaria— was convinced we had wrongly made what was historic prehistoric and, in fact, regarded the whole investigation of wetlands on both sides of the Yucatán Peninsula as contrived and the evidence as abstruse (Ochoa 1994; Siemens 1995a). The basic points have been in the literature for many years, but, clearly, the whole line of discussion has not been sufficiently well written nor, indeed, very well read.

Expansion of the Investigation

There are numerous wetlands around the Gulf of Mexico and across the base of the Yucatán Peninsula. It was obvious at the beginning that we needed to explore the extent of the patterning as soon as possible. We began with a speculative flight southwestward into the Usumacinta floodplain and the Selva Lacandona. There were tantalizing traces on several lake margins, but we have never been able to investigate them further on the ground (Siemens and Puleston 1972: 230–231).

Some hydrological considerations focused the search. The yearly regime of the Candelaria River is remarkably placid when compared to the larger trans-lowland rivers to the west, the Usumacinta and the Grijalva (Siemens 1978: 122). The first varies on the average only about one metre between high water and low water; the others about ten times that. On floodplains with volatile regimes, canalization and the buildup of planting platforms would have been difficult; on those with placid regimes it would have been much more feasible.

The Candelaria has a karstic catchment basin. It flows over terrain that is made up largely of limestone, which is subject to solution. In such basins a considerable fraction of the precipitation moves directly into the groundwater reservoir, reappearing

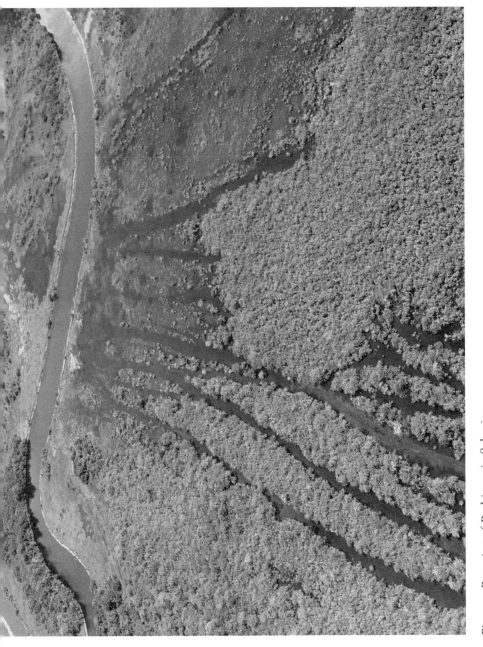

Fig. 1.5. Remains of Prehispanic fisheries.

gradually downstream through springs and seepage, keeping minimum discharge relatively high. Seasonal precipitation highs are
subdued by the retaining effect of labyrinthine subterranean reservoirs. The streams are relatively calm and reliable; the effect would
conceivably be enhanced by artificial dams. Since the floodplain of
one such stream had been found to be patterned with Prehispanic
remains, then others might have been similarly imprinted. There
were indications that the Hondo River and presumably also the
New River in Northern Belize were as placid as the Candelaria.

There were also indications of cultural historical relationships
between their basins. Sir Eric Thompson, who was still alive during our early investigations and keenly interested in these new directions that the study of Maya subsistence was taking, had traced
the migration routes of the Putún, also known as the Chontal Maya
(1970: 6). A major route led from the western to the eastern flank
of the peninsula. "Between A.D. 850 and A.D. 950 this once peripheral people controlled, in part or absolutely, northern Tabasco,
southern Campeche, Cozumel, Bakhalal and Chetumal along the
east coast of the peninsula" (1970: 4). We might therefore expect
that they would have recognized the potential of wetlands similar
to those of Acalan and used them similarly. Apart from the straws
thus thrown into the wind regarding the approximate chronology
of the field systems and the direction of the diffusion of the technology, these observations gave an edge to the hydrological indications and made an eastward extension of the search that much
more intriguing.

Northern Belize (figure 1.2, no. 2)

The New and Hondo wetlands proved, in fact, to be extensively
patterned (Siemens 1977, 1982a). We had ample basis for an expansion of the investigation when the continuation of the work in the
Candelaria basin was disallowed. The initial objectives were very
broad: we needed to enhance our understanding of the morphology
and ecology of ancient wetland agriculture, to gain some direct evidence of its function, to relate it to surrounding remains of settlement, and to clarify its chronology. We made considerable progress
on the first objective, less on the others.

Project participants made heroic forays from the main streams
into the wooded floodplains. On the basis of information from
these transects, as well as a fairly impressive array of vertical and
oblique air photographs, we hazarded a map of ancient wetland
agriculture in Northern Belize, some parts of which have held up

better than others. An arm of Pulltrouser Swamp, for example, that we lined with vestiges was later found to be quite devoid of them. We worked closely with the magnificent maps and descriptive material produced in the British Colonial Office (Wright et al. 1959) to make some distinctions in the environmental contexts within which the vestiges occurred.

Parallel to the investigations of the vestiges and their environmental contexts, some participants in the project undertook ethnographic inquiry into contemporary food production and consumption for the light that this might shed on ancient practice. A study of hunting and of the distribution of kill provided new insights into the present and the past significance of animal protein in the diet of the region, as well as the meaning of faunal remains in the archaeological context (Pohl 1985, 1990).

The archaeologists expended considerable effort and time on a series of cuts into ancient earthworks near the bank of the Hondo River at the village of San Antonio. They found some striking artifacts and drew intricate profiles. Unfortunately the location proved dubious as far as wetland agriculture was concerned; it is between the remains of a sizable Prehispanic settlement, as well as the present village of San Antonio, and the river. The possibilities for nonagricultural construction and perturbation were considerable.

Since those early efforts, a great deal of very valuable work had been done in Northern Belize by colleagues working on wetland agriculture at various locations (e.g., Turner and Harrison 1983; Scarborough 1983; Pohl 1985, 1990; Jacob and Hallmark 1996).

A sentence out of Puleston's report on the excavation near the Hondo River at San Antonio is interesting in very general terms regarding the chronology of wetland agriculture in this region and in lowland Mesoamerica generally: "All identifiable [ceramic] material is Early Classic [A.D. 400–600] and was apparently included with the fill used in later construction periods of the fields" (Puleston 1977: 38). Little creditable ceramic material was found in the excavations of fields in Pulltrouser Swamp, but the ceramics found in neighboring settlements extend over a period from the Late Preclassic, several hundred years before Christ, to the end of the Classic, at about A.D. 1100 (Fry 1983: 197, 211). The diagnostic sherds found underneath the remains of platforms located in the San Juan Basin of Central Veracruz, about which more will be said in chapter 2 indicate agriculture in these wetlands around A.D. 500 (Siemens et al. 1988: 107). There is thus the possibility of an approximate accordance, which certainly deserves further testing.

The Enigmatic Bajo (figure 1.2, no. 3)

The discovery of numerous vestiges of Prehispanic agriculture in the karstic wetlands on the two flanks of the Yucatán Peninsula inevitably stimulated speculation on the possibilities of finding similar evidence in the Prehispanic Lowland Maya core region at the base of the peninsula. Of particular interest were the numerous karstic depressions known as *bajos*. If intensive agriculture was to be regarded as basic to high cultural development, then one might expect to find vestiges of planting platforms and canals in the bajos around the major sites.

The reconnaissance of 1972 that yielded the patterning along the New and Hondo rivers also included the region of Quintana Roo westward and northward of Chetumal where one sees large shallow depressions. Several were closely patterned over vast areas, an order of magnitude more extensive than all the complexes of platforms and canals found in Northern Belize altogether (figure 1.6).

Puleston considered it likely that the patterning was the result of a natural soil-churning process often seen in regions of montmorillonitic clays, producing a pattern known as *gilgai;* he was skeptical of aerial reconnaissance without excavation being able to provide firm evidence for raised fields, particularly in the bajos (Puleston 1978: 234–236.) His skepticism is of some importance in that context, and of course is a very valid general caution, but, unfortunately, some readers took his statements to undermine visual evidence for wetland agriculture entirely, without making the necessary environmental distinctions (Ochoa 1994: 62).

The patterning in the depressions of Quintana Roo was eventually verified as artificial by Gliessman and his colleagues in 1980 (Gliessman et al. 1983). They concentrated on the Bajo de Morocoy and easily confirmed it to be a karstic depression, floored by clayey soils that become sticky and impermeable in the wet season, covered by semideciduous to deciduous vegetation and subject to periodic flooding—thus in many ways very similar to the bajos of the Petén. However, Morocoy was different in one key respect. A water table lay at 1.5 metres below the surface—at the height of the dry season. This is not characteristic of the bajos in the Petén.

The excavations undertaken in Morocoy showed an anthropogenic layer at something less than a metre below the remains of what were considered to be planting platforms, but not below the web of what were regarded as canals. The layer was taken as the original *bajo* surface into which the canals were cut and from

Fig. 1.6. Patterning in a *bajo* near Chetumal, Quintana Roo, taken during a commercial flight in 1960.

which material was taken to build up the platforms. This is parallel to what was found in the San Juan Basin of Central Veracruz. A dual planting regimen was hypothesized: the platforms planted in the wet season and the canals in the dry season. Gliessman and his colleagues did not recover direct evidence for cultivars and did not obtain any dates, but they did tentatively identify some ceramics from structures on "islands" in the bajo as pertaining to the Late Classic (A.D. 600–800) and some possibly to the Late Preclassic (880 B.C.–A.D. 150). This was similar to what Harrison and his colleagues had found in structures around Pulltrouser Swamp.

Researchers of the Instituto Nacional de Historia y Antropología (INAH) have begun a new study of the bajos northwest of Chetumal. This is being placed carefully into the context of previous work and includes a program of extensive transecting, coring, and test excavations. Also, the stratigraphic samples are to be split for parallel analyses at the laboratories of INAH and those of the Royal British Columbia Museum in Victoria, Canada. The substantive and methodological possibilities are considerable.

By the mid-1970s, in academic meetings and in many informal communications, those of us involved in the investigation of "raised" or "ridged" or even "drained" fields behaved rather like a fraternity of combat pilots painting trophies on their fuselages. The geography of ancient wetland agriculture in Mesoamerica still seemed incomplete; it was urgent that we extend the reconnaissance, certainly southward in the peninsula, into the homeland of the Classic Maya.

Bruce Dahlen and I crisscrossed the Petén of Northern Guatemala fairly systematically in March of 1976—from the mountains in the south to the northern border with Mexico. We paid especial attention to the bajos but did not find patterning of the sort found in the floodplains of the Candelaria River on the west and the New and Hondo rivers on the east (figure 1.7). It became apparent in a subsequent consideration of the karstic geomorphology and hydrology of the peninsula why this was not surprising (Siemens 1978).

The bajo is the *polje* of classic karst geomorphology, a steep-sided depression in a landscape underlain by calcareous parent material subject to solution. Water appears and disappears with an essentially seasonal rhythm, but the disappearance is often dramatically accelerated by drainage through perforations ("swallow holes") in the floors of the bajos and around their edges. The bajos of the Petén are, after all, perched several hundred metres above sea level. Retention and egress have been managed traditionally by

Fig. 1.7. A *bajo* in the Petén. One is tempted to read lines out of this vegetation; interpreters of SLR radar imagery of these features were also tempted. Neither has been convincing.

the Maya of the region. One can excavate and line depressions in the bajo floor for water storage (*aguadas*). Storage in the unmodified natural feature is not efficient or reliable. One can clear away rubbish from swallow holes and accelerate drainage if flooding menaces surrounding settlement. Being subject to both flooding and drought, the natural vegetational cover is both hydrophytic and xerophytic. The soils are clayey and vertisolic; that is, they alternately swell and crack with the coming and going of water. These are decidedly not the placid floodplain margin environments of the karstic rivers to the northeast and northwest, where open water can be found in the lowest parts of the backswamps at the height of the dry season and where the soils are enriched with regular additions of alluvium. Bajos would be difficult to cultivate, and indeed patterning characteristic of wetland agriculture has not yet been unequivocally demonstrated for such a context.

Why was settlement in some prehistoric epochs nevertheless arrayed around them? They may have constituted a natural defense, as at Tikal. They supplied water for domestic purposes, at least in the wet season, and were a source of fish. Transportation routes are commonly considered to have crossed the bajos, and some slight evidence was found for this during our reconnaissance, but not enough to set us to drawing maps of routes.

There is all this to the south, in the Classic Maya heartland, and then there is the Bajo de Morocoy to the north, patterned with the remains of planting platforms and intervening depressions. Instead of the several hundred metres of limestone, riddled with channels and caverns, that separate the seasonal, perched water tables of the bajos of the Petén and the basal phreatic horizon at approximately sea level, the bajo floors of Quintana Roo and Northern Belize are near sea level, as are the floodplains. The bajos also vary as to the nature of the sedimentation that has taken place within them. Permeability, seasonal drainage patterns, and vegetation may be expected to vary with these factors. The most useful single physical distinction may relate to soils. Vertisols dominate in the bajos of the Petén, as outlined, gleys or gleyed alluvial soils in the bajos and riverine wetlands on the eastern flanks of the Yucatán Peninsula. The clays in the two vary in composition; the clay content is lower and the silt content higher in the second. The first would have been more difficult to manage agriculturally with a Prehispanic technology than the second (Darch 1980: 2.6.20;).

It is quite possible that the natural conditions of the Petén were not always what they are today. One might want to invoke climatic change, but various implications of the expansion and intensification of cultivation during the Maya ascendancy also suggest themselves. Erosion in the hill land and resultant silting in the bajos may well have enhanced the lamination of bajo floors and resulted in more efficient water storage. Conversely, one may deduce that the length and level of bajo flooding decreased with reforestation and the reduction of erosion consequent on the collapse.

An early observer forecast that "the exploration of [these] uninviting yet interesting swamps will repay in discoveries all effort devoted to them" (Lundell 1937: 29). They are notoriously difficult to penetrate on the ground. Water behaves capriciously within them and their stratigraphy is in gentle shades of gray. They have flirted with airborne observers, suggesting patterning to some but not to others, in some locations and not in others. To the same degree, of course, they remain intriguing. For a time it was thought that the riddle had been solved.

Radar.   In the late 1970s a rumor coursed among the cognoscenti that remarkable new evidence was about to show raised fields to have been very much more extensive in the Maya lowlands than anyone had imagined. The truth was eventually revealed in the media, then followed by the real truth published in a prestigious context.

On June 3, 1980, John Noble Wilford reported in the *New York Times* that, "An airborne radar survey over the dense rain forest of Guatemala has led archaeologists to the discovery of an elaborate network of ancient canals." This was a relatively careful and indeed a crafty piece. One of the savants of Mesoamerican prehistory was interviewed and allowed to draw the spectacular conclusions regarding Mayan subsistence. Other publications were less inhibited. The *Capital Times* of Washington headlined its story: "Mayans used canals to avert food crisis: radar" (June 3, 1980: 5). *Newsweek* made it simple: "The Mayas' Secret: A Canal Network," and went on to explain how the "Mayanologists" had found the answer to the great mystery of Maya subsistence: canals had made swampy wastes cultivable (June 6, 1980: 24).

All this was followed by a serious and indeed very careful piece by the discoverers themselves in *Science* about "Radar Mapping, Archaeology, and Ancient Maya Land Use," inferring much but also indicating clearly that much was still very tentative (Adams et al. 1981: 1457–1463). It was not read as carefully as it was written. What might be so was taken as established and many of the technical qualifications just simply escaped notice.

The fate of a photograph and a tracing from it are diagnostic of the fundamental problems that still cloud this material. The photograph was an ordinary air oblique that I had taken over a part of the floodplain of the Candelaria with a handheld Rolleiflex from the open door of a Cessna (Siemens and Puleston 1972: 234). In the 1981 publication on the radar evidence, a tracing from that photograph was presented, showing what were taken to be vestiges of canals visible in the portion of the floodplain shown. It was captioned: "Overlay of a low-level oblique aerial photograph showing confirmed ancient canal patterns in the Río Candelaria zone of the Maya lowlands. Scale about 1 : 1000" (Adams et al. 1981: 1461). It was obviously included in order to provide a corroborating comparison with tracings from the radar imagery. It clearly implied that here was the way to read a wetland.

A floodplain is normally as veined as a retina due to the vagaries of the flow of water, sedimentation, and plant colonization, as well as human incursion. It presents a configuration out of which one can read what one wants unless inhibited by ground information and analogues. The tracing that was made is difficult to credit. It shows numerous lattices, of the sort also seen on the radar imagery. Unfortunately the scale of the lattices seen on the photograph indicates highly improbable planting platforms in the order of a few square metres. This and other aspects of the morphology of the platforms and canals are completely contrary to anything that had been actually found and investigated on the ground or from the air on the two sides of the Yucatán Peninsula up to that time and published in various places. The implied comparability between what was derived from a photograph taken to be of an approximate scale of 1 : 1000 with radar imagery of a scale of 1 : 250,000 is, of course, in itself highly questionable. Moreover, the two complexes of vestiges of planting platforms and canals that are clearly apparent in the photograph, which had in fact been confirmed on the ground and by numerous analogues, were not recognized as such. There was an additional unfortunate touch; canals had been traced out in a field on undulating terra firma. Here was an image that had been misread indeed.

The misrepresentation was compounded not long afterward in a very substantial review of Lowland Maya Archaeology in *American Antiquity* (Marcus 1983: 474). The tracing referred to was presented again, accurately redrawn but not so accurately recaptioned: "Apparent canal systems along the Río Candelaria, Campeche, Mexico as detected by radar (redrawn from Adams et al. [1981:1461])." I gained a new respect for my trusty Rolleiflex; I had never suspected it had radar capability. In the text of the article, the tracing was cited as a confirmation: "The few ground-truth checks made so far do confirm the existence of the fields and canals." The drawing unfortunately does not represent a "ground-truth check" nor does it confirm anything; in fact it obscures what was found and checked on the ground.

Some of the simpler aspects of the radar imagery, which affect visual interpretation, need to be reviewed. Firstly, the imagery has a scale of 1:250,000, as noted. Then there is the question of texture: the imagery is extremely grainy. Under those conditions a feature needs to be at least 20 metres wide and 100 metres long before it can be resolved as a waterway (Pope and Dahlin 1989: 95). The Candelaria River itself just manages to leave an unequivocal line. Only substantial canals are visible; the web of canalization that typically defines a wetland agricultural complex cannot be seen.

The proponents of the radar evidence were admirably open about all of this. Various interested investigators, including this author, were invited to a review of the radar imagery in the Jet Propulsion Laboratory in Pasadena in August of 1981 and participated in the "blind experiment" cited in a rebuttal to a critique of the radar evidence by Pope and Dahlin (Adams et al. 1990: 242). We were asked to trace what we saw in Pulltrouser Swamp. In my journal for that day I had noted about that test and our review of other imagery from over the wetlands of the Yucatán Peninsula that "I can see lines but I wouldn't always draw them as he (Adams) does."

I subsequently scanned copies of all the relevant radar imagery, which was very kindly made available to us by the Jet Propulsion Laboratory, and undertook a careful scrutiny of the lowland, wetland portions. I could identify some lines in several places and match them with prominent canal vestiges that I had seen on my own oblique air photography. These were large canals, approximately of the width of the rivers themselves, but I could not verify anything like the web shown on the startling map of canals in Northern Belize derived from radar that, perhaps more than any other single figure, has dramatized this line of evidence in the minds of a wide readership (Adams et al. 1981: fig. 4, 1460). I certainly saw nothing like the simplified web published in the *New York Times* and elsewhere in the popular press.

The large canals will have served various hydrological or transportational purposes, but agriculture cannot be deduced directly from them. It seems hardly justifiable to make a statement such as the following:

The recent radar mapping discovery of widely distributed patterns of intensive agriculture in the southern Maya lowlands provides new perspectives on classic Maya civilization. Swamps seem to have been drained, modified, and intensely cultivated in a large number of zones. (Adams 1980: 206)

The unwarranted impression of direct radar evidence for agriculture aside, the indicated function of the canals is simplistic. It was not just drainage but also and importantly storage. The whole system that was in operation in the wetlands and surrounding hill land of the Maya lowlands and other lowlands to the west is still not often referred to carefully in the literature. The "raised field" does not deserve to be the cliché that it has become. And, anyway, the initial discovery was made in 1968 (Siemens and Puleston 1972).

This whole controversy, as already indicated, was courageously taken on by Pope and Dahlin (1989: 87–106). Their scrutiny of the actual imagery and related evidence has been diligent indeed, and one must concur in what seems their basic finding, from which a great deal proceeds: "The spatial resolution of, and speckle noise in, existing radar imagery make it inadequate for mapping lattice patterns of small canals" (1989: 87). They develop the various implications well: webs of small canals, the direct evidence for wetland agriculture, have not yet been verified satisfactorily for the bajos of the south-central Petén. They conclude that "the majority of densely-populated Classic Maya sites were not dependent on wetland agriculture" (1989: 87). It is necessary to reiterate a qualification in order to counter simplistic treatment of Maya subsistence: its ingenuity is likely to have lain in the orchestration of various productive activities. The discovery of wetland agriculture in itself, although it opened a new perspective, had best not be seen as the solution of a problem—and radar is not yet a Rosetta stone.

It may be possible eventually to penetrate vegetation coverage by remote sensing at an adequate scale and with a sufficiently tolerable margin of vertical error to read the surface under the forest for vestiges of human activity that are measurable in a few metres. This would vastly enhance the understanding of larger patterns of settlement, transportation, and land use. Of course, as deforestation and the expansion of ranching proceed apace, we are obtaining more and more out of the visible portion of the spectrum.

In the meantime, a "new orthodoxy" seems to prevail among the students of the Maya (Pope and Dahlin 1989: 89; Turner 1993). The vast network of bajos in the ancient Maya core area, previously seen as interstitial and certainly as enigmatic, is now taken to have been the very basis of food production for at least the Late Classic Period. This network, radar, and the solution of the subsistence problem seem fairly durably associated. One could cite a variety of evidence. Recently Ronald Wright, a prominent travel writer who had given every indication of having done a great deal of reading in preparation for his earlier travel, made this rather sweeping statement:

> Until about a dozen years ago archaeologists were baffled by the apparent lack of an economic base for Maya cities. The breakthrough came when satellite and airborne radar pictures revealed networks of ancient ditches and canals crisscrossing enormous swampy basins—the Classic Maya had converted these wetlands into something like the famous "floating gardens" of Aztec Mexico. The canals were designed to hold water from the wet season through the dry. The fields between them were raised above water level and kept fertile with muck dredged from the canals, vegetable mulch, and by "self-manuring" with human and household waste. These methods allowed large increases in food production but may have made the Maya vulnerable to climatic disturbance, plant disease, parasites, and social unrest. (Wright 1989: 54)

This is a misrepresentation on a number of grounds: The "breakthrough" already had been in the literature many years earlier. The model of the system that is assumed is highly questionable; such wetland agriculture as has been verified in the Maya lowlands had best not be considered as chinampa agriculture without qualification and certainly not as "floating gardens," which is one of the oldest, unfounded misconceptions about this sort of agriculture in Central Mexico. The storage function of the interstitial canals, however, is recognized and expressed more clearly than in many academic treatments of this system. And to repeat: such canals as are evident on the radar imagery are not direct evidence for wetland agriculture nor can it be implied from them.

There is little that can now be done about the misrepresentation. The affirmation of this great new thing was too forcefully made, in the media and in the prestigious academic journals.

"Empty" Wetlands (figure 1.2, no. 4)

Lowlands continue westward from the Yucatán Peninsula, around the southern arc of the Gulf. *Lowland* is actually a misnomer in this instance, except in general terms, since this terrain has a good deal of geomorphological variation and topographic relief. It is interspersed with many wetlands, which have yielded very little patterning—only in that sense are they empty.

We carried out numerous reconnaissance flights over these lowlands between 1960 and 1968—mostly in aid of an examination of the rather remarkable mid-twentieth-century expansion of agricultural settlement and ranching within them, Mexico's "march to the sea." We had learned to appreciate the difference between the landforms developed in these and other tropical lowlands on Tertiary and Quaternary sediments, between the hill land of the "lowlands" and the real lowlands, largely subject to inundation. The expansion of settlement and ranching in the 1960s was largely a hill-land phenomenon; it was proceeding rapidly from a series of long-settled cores and had a crenelated forward edge which it seemed reasonable then to call a "frontier." We took photographs of wetlands but paid little attention to them. There was no stimulus yet to search through them for the remains of ancient field systems.

After the patterning had been recognized in the wetlands of Central Veracruz in 1976, the photographs that had resulted from the pre-1968 flights were scrutinized again, together with relevant runs of commercially available vertical air photography. Also, a new reconnaissance flight was undertaken from Jalapa, Veracruz, to Villahermosa, Tabasco. It seemed important to reexamine the vast interstice over which we had launched ourselves, between the patterned wetlands of southwestern Campeche and those of Central Veracruz.

The flights included the surroundings of the Usumacinta River, from the Lacandon region of eastern Chiapas northward into lowland Tabasco. Several wetlands were patterned, as already indicated, one of them in a very precisely rectilinear fashion. However, they were difficult to reach and have not yet been investigated on the ground. We repeatedly overflew the lower Mezcalapa (Grijalva) River. It is a world of endless convolutions, fairly dense settlement, and an intricate patchwork of fields and pastures—microaltitudinally differentiated. We detected a few instances of canalization and linear mounding, but nothing like we had seen in Campeche or Northern Belize. We searched the lower Coatzacoalcos-Uxpanapa system—the lowland described magnificently by Coe and Diehl (1980), now under a pall of petrochemical pollution. Of particular interest were the extensive grasslands around the San Juan River of Southern Veracruz, pitted with many shallow lakes but with no sign of patterning. The labyrinthine lower Papaloapan Basin was similarly unrewarding.

These vast floodplains are worlds of subtle variety. The mean annual precipitation in the Southern Gulf Lowlands exceeds 2000 millimeters; it falls off to the east and west. This is definitely a humid tropical environment, but there is a dry season of several months duration. Some means of cropping during the dry season will have been advantageous in Prehispanic times, as it is now— more in some years than in others. Total precipitation may vary considerably from year to year. The lowlands are also affected by variations in the precipitation over the largely impermeable catchment basins of the larger rivers in the mountainous country to the south. All this induces considerable variability in lowland flooding.

The natural vegetation of the lowlands includes a complex array of beach and levee formations as well as rain forest and succession species on higher ground. The wetlands proper may be occupied by various types of mangrove or shrub forest dominated by palm. However, the principal swamp vegetation, at least from the lower Usumacinta to the lower Coatzacoalcos, is *popal*, a community of herbaceous aquatic plants with various patterns of dominance. These areas are flooded to various depths for most of each year, but they may dry out just sufficiently to make pasturing or even cultivation feasible, as will be seen.

The rise and fall in the levels of the major lowland rivers, the prime variable in the lowland environment, is exemplified by a twenty-five year run (January 1945–December 1969) of river-level measurements from Samaria, a hydrometric station on a distributary of the Mezcalapa (Lower Grijalva) River, forty kilometres west of Villahermosa (SRH 1971a). The figures show the pronounced seasonal and diurnal fluctuation that is characteristic of all of the major lowland channels. The difference between the extreme maximum and minimum was 6.34 metres. Yearly differences were commonly around 5 metres. The Mezcalapa was traditionally feared as the most menacing stream of Tabasco. It has changed its course repeatedly in historic times. The Papaloapan also frequently devastated its surroundings before it was dammed in 1954. All in all, therefore, the hydrography of the main river floodplains of the Southern Gulf Lowlands seems rather volatile for wetland agricultural systems such as those indicated by the vestiges found in Campeche, Northern Belize, and neighboring Quintana Roo. However, a succession of agriculturally very useful micro-environments is seasonally laid bare by the fluctuating streams.

In the lowlands there are traditional ways of complementing wet-season agriculture on firm ground with dry-season cropping on seasonally exposed terrain; we may expect that something similar was done in Prehispanic times. Coe and Diehl have described dramatic yearly variations in water levels and corresponding land use in their impressive study of present subsistence around the large Olmec site of San Lorenzo Tenochtitlán, near the Coatzacoalcos River, and have also drawn the Prehispanic analogy (1980). West and his colleagues had already found similar fluctuations in the Tabasco lowlands (1969: 67). Toward the end of the dry season, in April or May, when the water levels are at their lowest, one can see a succession of land use zones along the backslope of many a levee. Farmers and ranchers have long followed the seasonal descent of floodwaters along this incline with their fields and herds. North of Villahermosa is a region of line villages, with narrow properties that transect such series. Evidently the engineers subdividing old hacienda lands during the process of agrarian reform in the 1930s cut the parcels to be given to the peasants in such a way that each had access to terrain at all levels of the profile. In one way or another farmers needed to obtain access to a combination of microenvironments; subsistence was difficult in these lowlands without it. Much of all this has been disturbed or abandoned in recent years as a result of the impact of the exploitation of petroleum, including extensive pollution, as well as the move toward industrial employment and into the city, to say nothing of the overall economic impact of globalization.

There is no reason to believe that the prehistoric inhabitants of the Southern Gulf Lowlands were not aware of techniques developed to the east and west of them for the modification and more intensive use of swampy terrain. A major east-west overland trade route crossed the Southern Lowlands before contact with Europeans (West 1969: 103). A system of planting platforms and canals was probably not only difficult to build and maintain in the hydrological context of these lowlands but unnecessary as well. There would have been other ways of complementing the wet-season yields of higher ground with dry-season crops—by exploiting the rivers' regimes as they were.

From the seasonal progression of activities along the gently sloping margins of wetlands in these volatile floodplains and what may be seen around wetlands elsewhere in the tropics, one gains some impressions of what seasonal, catch-as-catch-can agriculture was probably like on the floodplains of the more placid rivers, as well, before intensification became necessary and canalization was begun. Subsequent modifications of the surface and manipulations of the water secured and enhanced it. One may envisage a continuum.

It is perhaps ironic that recent experimentation in the intensification of wetland agriculture has been taken further in the Southern Gulf Lowlands, where there is little evidence for planting platforms and canals, than in the Maya lowlands or in Central Veracruz. I refer to the Camellones Chontales northwest of Villahermosa—a bold rectilinear incursion indeed when seen from air. They were built by engineers with modern equipment and represent imaginative transfers from the chinampa system of Central Mexico rather than deductions from the vestiges of lowland systems. There have been serious technical and social problems, but recent indications are that this system has come to provide significant sustenance for the community involved.

Northern Veracruz, the "Huasteca" (figure 1.2, no. 5)

Before we committed ourselves to a detailed investigation of the patterning in the San Juan Basin, we explored the Gulf Lowlands northward of the mountain spur that interrupts them, about thirty kilometres north of Veracruz. We undertook three flights across the lowest portions of the floodplains of a whole series of streams, from the Misantla to the Panuco, and one exploratory foray on the ground. The initial objective was simply to see if wetland patterning would show up in those many locations that were roughly analogous in their topography and hydrography to the patterned basins of Central Veracruz and the Maya lowlands.

We also wanted to follow up on various intriguing indications about this region. Peter Schmidt had described the dramatic patterning he found—by chance—just north of the Nautla River (1977). Jeffrey Wilkerson had published photographs of what he considered a network of canals in the floodplain of the Tecolutla River (1980). José Melgarejo Vivanco had reviewed legends and ethnohistory in various early Spanish chronicles and concluded that the origin of the chinampa lay in the surroundings of Tamiahua, not far north of Túxpan (1980). In the background was G. F. Eckholm's contention that there had been links between the Maya lowlands and Northern Veracruz (1944).

Various features have intrigued us in the Huasteca and invite investigation. In the surroundings of Tamiahua, for example, there are long straight lines across unpatterned floodplains, much as we saw them on the flanks of the Yucatán Peninsula. One may again deduce a transportation infrastructure, channels that are useful in order to get from streams to higher agricultural land beyond the floodplain and to escape or surround waterborne invaders. It may be, also, that these are the remains of ancient fisheries.

Near the mouth of the Túxpan River, already within saltwater marsh, there is an "island" stepped with terraces and linked to the river channel by a straight waterway that looks very much like a canal. We circled over this site repeatedly but never had the opportunity to search for what might already be known about it or to visit it on the ground. It could make someone an excellent thesis!

We also found numerous remains of planting platforms and canals (Siemens 1982b, 1983a). They appeared predominantly as pods on the backslopes of levees and sometimes across backswamps along the lower courses of the ladder of streams that cuts the Huasteca. A characteristic assemblage of features and environmental relationships recurs. From it one can deduce an agricultural strategy that came down, in large part, to the securing of dry-season cultivation and with this the reduction of risk in the overall subsistence systems of lowland communities.

One particular combination of landscape features past and present may be seen again and again. On levee tops one often finds remains of Prehispanic settlement. The large site of Santa Elena on the Estero de Tres Bocas north of the Nautla River is a fine example (figure 1.8). This is also where central facilities of ranches, other current settlement, and plantations of commercial crops are to be found. A kind of cultural and hydrological strand-line divides the reasonably secure levee tops from the lower land subject to seasonal inundation. The plantations give way in this general area to pasture, and the Prehispanic canalization becomes apparent. The patterning is, at first, predominantly downslope, and then both downslope and transverse. In the lowest areas there is often unpatterned swampland, with open water in places even at the height of the dry season. What was seen in the wetlands of the Huasteca thus contributed markedly to the conceptualization of wetland agriculture in general.

Summary: A Heuristic Model

During the protracted search and the San Juan Basin investigation that is to be discussed in the next chapter, we were gradually able to clarify what linked the various examples of patterning. They

Fig. 1.8. A typical series of features, past and present, along the Estero de Tres Bocas, not far north of Nautla: prehistoric site and modern ranch facilities on the levee, remains of "raised fields" and modern pastures farther down, swamp in the lowest areas.

have a typical arrangement and a typical topographic, geomorphological, and hydrological context, all of which can be simply represented along a gentle slope (figure 1.9). One can also make various generalizations regarding function and change. The construct has proved comfortable in many a discussion of wetlands.

Prehistoric wetland agriculture was the initial concern, but more and more it became apparent that historic and, indeed, present wetland land use as well, could often be placed to advantage along that same generalized slope. It became a raked stage for the entire book.

We need to emphasize the lowland floodplains with dampened hydrological regimes. This distinguishes between floodplains with through-flowing streams coming out of tributary regions well up in the Sierras, and the floodplains surrounding the affluents of these main streams, joining them near their mouths, as well as the floodplains of rivers with karstic basins. However, with some manipulation of the angle of the main line, the model can accommodate the first of these too, as well as highland lake margins and thus, in fact, most tropical wetland margins.

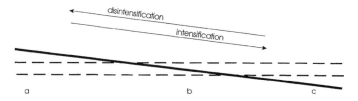

Fig. 1.9. Heuristic model of wetland margin, with terrain: (*a*) above the normal level of inundation; (*b*) between yearly maximum and minimum water levels (ecotone); (*c*) saturated or flooded all year. Arrows show the directions of possible change in land use over time.

The model takes into account the seasonal rise and fall of water levels, the metronome that beats behind any discussion of tropical floodplains. Levels have been routinely measured in the channels of the larger through-streams, but not usually in their tributaries. Information on the latter was elicited by field inquiry from the first forays onward: How high does the water normally come? Up to about the waist. How low does it drop? Down to about so much in a posthole. Later we were able to watch one or two floods and sharpen our inquiry. In the background were the accounts of the dramatic rhythm of water levels, agriculture, and ranching on the floodplains of the Amazon lowlands (Hiraoka 1985; Roosevelt 1980).

Landscape Components and Associated Uses

The model has three main elements, or zones: (*a*) terrain that is normally safe from inundation but does not benefit from it either; (*b*) that which is between mean high water and mean low water— the ecotone; and (*c*), that which is taken up by swamps where water is just above or just below the surface in even the driest period, or by lakes.

Terra firma (zone *a*) can take two main forms. On the one hand are the Tertiary sediments, just landward of the floodplain or on islands within it, with sandy, clayey soil that is quite compacted and may be cemented at a certain depth. The remaining natural vegetation is likely to be low tropical deciduous forest (*selva baja caducifolia*). In Mexico such land is held in *ejidos* or private properties of various sizes, its main uses are shifting cultivation or wet-season pasture. With the provision of canal or spray irrigation, the use of the lower sections of this terrain can be intensified to commercial agriculture or improved pasture. On the other hand are the higher reaches of natural levees that normally do not flood. This recent alluvium is topped by mostly well drained soils that remain moist by reason of seepage during the dry season—prime agricultural land. Both levee tops and the higher, safe margins landward of the floodplain serve as settlement sites, now as in the past.

At the base of the model, zone *c*, is the more or less open water of swamps or lakes, thick with patches of hydrophytes and forest. It is easily set aside, a perceptual blank. It was not a blank for the ancient inhabitants, but rather an area in which to obtain important supplements to agriculture by hunting and fishing. It draws hunters still, but the tendency has been to drain land like this and thus to draw down the boundary between zones *c* and *b*.

Zone *b* is the zone of greatest interest here. It is an ecotone, a term that has been used mainly with respect to the ranges of plant and animal species and thus must be regarded as a metaphor in the discussion of human land use. It is considered a zone of transition between two communities; within it various *seres*, or processes of biotic succession are in juxtaposition. It is attractive to a variety of living forms, which leads toward a greater variety and density of species than in the communities flanking it—the "edge effect" (Odum 1971: 157–158). However, it is not simply a boundary or an edge; there are mechanisms within it that do not exist in the juxtaposed ecosystems (Décamps and Naiman 1990: 2). It is dominated and indeed enlivened by the pulse of flooding (Junk, Bayley, and Sparks 1989).

The boundary between zones *a* and *b*, particularly when it is the boundary between Tertiary and Quaternary sediments, typically shows up as a kind of strand line expressed in the "language" of vegetation, i.e., tonal variations on air photography—lighter tones above, darker tones below. The line is extremely tortuous and is often lost under the higher vegetation. In our air reconnaissance we have focused repeatedly on the areas just below this boundary because that is where the patterning is to be found.

Left to itself zone *b* is likely to be covered in forest, probably *selva mediana subperennifolia* (that is, a tropical forest of medium height); human perturbation leaves only certain succession species. The soils are dark and clayey; they crack extensively during the dry season and swell again with the rains. This terrain was and is of great interest to agriculturalists who are able to grow dry-season crops on it and thus supplement their agriculture on the neighboring hill land in critical ways. In prehistoric times wetland agriculture here was intensified by the cutting of canals and the building up of planting platforms. Not much of this terrain is accessible to agriculturalists now, since it is largely in the hands of ranchers, as it has been since early colonial times. They consider it of central importance to their operations and try more and more to secure and enhance their use of it through individual drainage and irrigation projects. At various times governmental agencies have carried out more ambitious hydrological projects—affecting all three zones.

Variations (figure 1.10)

The slope can be oriented in any direction, of course. As it follows river curves, particularly meanders, it may occlude within the *b* zone (figure 1.10 [2]), leaving no swamps or lakes but only a trunkal drainage channel and a wide expanse of terrain that seems to be an optimal situation for the development of large complexes of planting platforms. Several examples out of the San Juan Basin will be discussed in chapter 2. In an attenuated floodplain, by contrast, as along the Candelaria River or one or another of the rivers of the Huasteca, the complexes are usually of more limited extent and strung out along the landward margins of the natural levees (figure 1.10 [1]).

The gentle slope steepens in the floodplains of the large through-flowing rivers—the Usumacinta, the Grijalva, the Coatzacoalcos, and the Papaloapan—but maintains its three main subdivisions (figure 1.10 [3]). Along the Grijalva, for example, even a brief investigation with West and his colleagues' work in hand (West et al. 1969) shows an ample vertical sequence of phased cropping opportunities within zone *b*, tuned to the substantial rise and fall of water levels within the vast and convoluted floodplain. Intensification by means of canalization and platform buildup seems to have been unnecessary or impractical.

Other lesser variations in slope and accidents of sedimentation may be identified. Generally, however, the scheme fits and clarifies most instances of patterning from Northern Veracruz to Northern Belize.

The range of uses that may be envisaged for zone *b* (figures 1.9 and 1.10) are bracketed on the one extreme by what can be called fugitive or flood-recessional agriculture, and on the other by the chinampa. An intriguing instance of the first was found on a wetland margin near El Palmar, a community just northwest of the San Juan Basin (figure 1.11). Not long after the discovery of the Central Veracruzan wetland patterning, some colleagues and I were attempting to clarify the microtopography and vegetation associated with a particularly interesting complex seen from the air. Dense cane grass two metres high barred the way into the wetland; our attention fell by default on the mosaic of small fields behind us in which

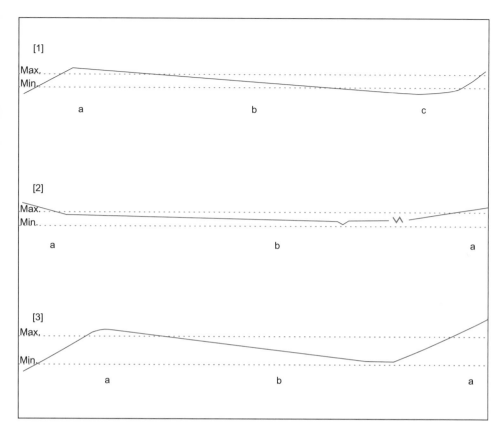

Fig. 1.10. Variations in the topographic expression of the elements of the model shown in figure 1.9.

dry-season agriculture was being carried out just above low-water level and subject to wet-season inundation. We have been able to observe this terrain repeatedly over a period of years and obtain fairly detailed information from local informants regarding current agricultural practice, all of which has provided useful points of departure for the interpretation of ancient practice (Siemens 1989a). Other investigators have since examined the history and agricultural economy of the community as a whole (e.g., González Jácome n.d.); it fascinates us still and serves as something of a touchstone.

Fugitive agriculture involves an exploitation of surfaces exposed by low water with quickly maturing cultigens, as possible and necessary, from year to year, without any attempts to manipulate water levels and planting surfaces through canalization and buildup.

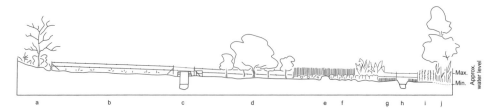

Fig. 1.11. Profile of an actual wetland margin near El Palmar during
the dry season: (*a*) *Selva baja caducifolia,* known commonly as *monte;*
(*b*) wet-season *milpa* (maize field), with detritus of harvested crop; (*c*) a
lined well, enclosed in a fence with a movable water tank; (*d*) fruit trees;
(*e*) *plantél,* or raised starting bed; (*f*) a range of crops in the zone of inundation, known as *bajío* or *orilla;* (*g*) introduced pasture grasses; (*h*) intake
for motorized irrigation system and water source for cattle; (*i*) boldest of
the dry-season *milpas;* (*j*) edge of the swamp, with high forest species and
hydrophytes intermixed.

The final phase of such a venture can be quite risky, when an early
onset of the rains can easily inundate a crop. It is always important
to get the harvest out on time. Such activity can also, more prosai-
cally, be called flood-recessional agriculture. Wilk's work on such
agriculture in Belize (1981) and the investigations of the *marceño*
in Tabasco by Orozco-Segovia and Gliessman (1979) have been particularly useful. My own work on agricultural forays at low water
into artificial water catchment depressions, the *albarradas* of arid
southwestern Ecuador, has added nuance (Siemens 1981).

One can think of this as a distendable or intensifiable portion of
a community's subsistence repertoire, a way of reducing risk. In it
one can see the antecedent of, or alternative to, the system that
left the patterning in the lowlands.

On the other extreme is the well-known chinampa as seen in its
major surviving complexes south of Mexico City—so easy to get
to and, after penetrating the tourist spectacle on the periphery,
such a pleasant labyrinth in which to spend an afternoon. These
complexes are rapidly being built over, and the water in the canals
is heavily polluted. Crops are grown on what remains of the built-
up and carefully leveled platforms, separated by canals. Water levels are controlled and flooding is prevented. When the metropolis
had not yet captured the outflow of springs at the base of the uplands to the south, a continuous supply of fresh water was fed in.
The system has been extensively studied, including the inquiry of
Parsons et al. into its chronology (1982), Niederberger's investigations of the early exploitation of lake margins in the Basin (1979),

Venegas's study of one complex of chinampas, those of Mixquic (1978), Jiménez-Osornio and Gómez-Pompa's synthesis (1987), A. J. Robertson's analysis of this agriculture as a resource system (1983), and a great deal more. Often this system has been taken without much reflection as an analogue for the agricultural remains found in the various wetlands discussed in this chapter. It is seen here as an elaborated version of most of the wetland agriculture in the lowlands.

Between these extremes are the structures that are central to the study of Prehispanic wetland agriculture in lowland Central Veracruz and the plane of departure for the consideration of subsequent developments in the wetlands. A designation such as *proto-chinampa* may be appropriate. This term implies a process of development, which may well have obtained in general terms, not so much in any given complex. This term emphasizes that the structures are not to be taken uncritically as chinampas. Other terms, such as *ridged, raised,* or *drained fields,* have been used, each implying preconceived notions of structure and function that are not mutually exclusive. Best perhaps to avoid all this and use something neutral, such as *platforms and canals.*

Dynamics along the Gentle Heuristic Slope

Long-term changes in the physical environmental context of wetland agriculture, particularly changes in the relationships of floodplains to sea levels and in precipitation, can be expected to have affected the wetland agriculture fundamentally, making it possible and then again eventually impossible in any given situation, or vice versa. One may envisage, for example, a drop in the seasonal amplitude between high and low water as facilitating an incursion and a rise as precluding it, or conversely a rise that makes wetland agriculture possible in areas where it was not feasible before. The sedimentary and related vegetational evidence is difficult to decipher; the effects of tectonic activity *vis-a-vis* climatic change may well be hard to distinguish along any given coastline. Ongoing paleoecological work by various investigators promises some eventual clarification (Gunn, Folan, and Robichaux 1995; Hodell, Curtis, and Brenner 1995). Some observations made from the air in this regard, just after a substantial flood, will be reported in chapter 2.

A further natural change is normal in wetlands, particularly their margins. They are constantly being filled in by siltation. Aquatic plants resist water movement and hence facilitate deposition; dead plant remains add to the accumulation and gradually irregular hummocks appear above the surface. This is accompanied by a plant succession, a *hydrosere,* that begins with aquatic vegetation and ends in tree cover. Planting platforms may quite possibly have originated as a logical acceleration and entrainment of the process by which a wetland naturally fills itself in and clots into amorphous islands.

Keeping natural conditions generally stable, one can envisage changes in communities' investment of labor and means in the agriculture of the wetlands. Two arrows can thus be laid along the gentle slope (figure 1.9), pointing in opposite directions. The one pointing downslope can be taken to represent intensification. It begins with fugitive exploitation, which is secured, as needed, with progressive canalization and build-up, beginning with the longer parallel downslope canals, the remains of which are common in many places, and then continuing down toward the swamps with a quadrate webbing. In restricted pockets, diking and the induction of fresh water may have been possible, allowing control of water levels and year-round cultivation. In these circumstances, yes, one can speak of chinampas, but in the lowlands these seem to have been the exceptions.

These developments may be seen as a progression from rudimentary to more sophisticated technology. But what we see may be various adaptations, all ingenious and effective. Very simple facilities like a hand-dug canal or a quickly heaped up temporary dam can achieve subtle and highly beneficial effects. Better perhaps to avoid the rankings and speak of elaboration.

When populations decline or relocate, when priorities change, when conflict sets in, the process may be reversed. This may be represented by an arrow that moves in the opposite direction. Maintenance lapses, perhaps first in the more demanding lower areas, and then upslope; the platforms are soon rounded and the canals filled in. Fugitive agriculture may persist sporadically and then lapse as well. Reforestation sets in, and one can imagine that under tropical conditions all remains are covered over in a few years.

If we envisage concurrent natural and cultural change, vigorous adaptations could counter natural change, such as intensification of wetland use in response to increasing aridity and dropping yields on terra firma. Such a period of drying is now fairly clear for the Maya region in A.D. 800–1,000 (Hodell et al. 1995). On the other hand, a disintensification already fostered by social stress, say a lowered capacity for the maintenance of public hydrological works, could be accelerated by natural change.

2. Deducing the San Juan Basin in A.D. 500

In such a place there could be no hunger?

—after Durán as cited by Melgarejo 1980: 120

Approaching the Basin along Highway 140 just west of Veracruz, even with a topographic map in hand, one must watch the dips and rises carefully or one will miss it. At one point in the road one can catch a glimpse of its bottomlands, with just a hint of patterning, palms in the foreground and dunes in the distance (Frontispiece). From the air it is even less apparent, without some very careful map reading, unless the rains have been heavy and the central lake, Laguna Catarina, has expanded to demarcate the bottomlands (figure 2.1). A basin is only really clear when the topography is exaggerated (figure 2.2).

Cartographers have had difficulties with the San Juan River; its channel is obscured by the hydraulic works to be discussed in chapter 7, as well as by seasonal flooding. The inhabitants are mostly convinced that it exists. It gathers up the streams of the shallow depressions and drains into the La Antigua River just opposite the town of the same name.

The Basin is elusive in the documents as well. Not all, but most colonial travelers, from Cortés onward, pass just around the Basin to the north or south and say hardly a useful direct word about it. Travelers who come this way in the nineteenth century note a low forested area, register their concern about the miasmas that must be emanating from the swamps within it, and then pass quickly on. One would have thought that the engineers of the Secretaría de Recursos Hidráulicos (SRH) planning a massive intervention in the basin would have been able to present it clearly, but unfortunately their physical environmental and land-use data are aggregated largely according to the basic political entities, the *municipios*, which certainly obscures microenvironmental boundaries and ecological considerations. This is a general problem in the interpretation of all sorts of Mexican national statistics.

Fig. 2.1. Aerial view of a part of the flooded San Juan Basin. The channels of the San Juan River and a tributary, straightened by the Secretaría de Recursos Hidráulicos, cross from bottom left to top right. A fragment of the preintervention hydrography shows through on the bottom middle. An "island" of older sediments in the bottom right bears a corral and milking facility (*ordeña*), probably also the remains of Prehispanic settlement.

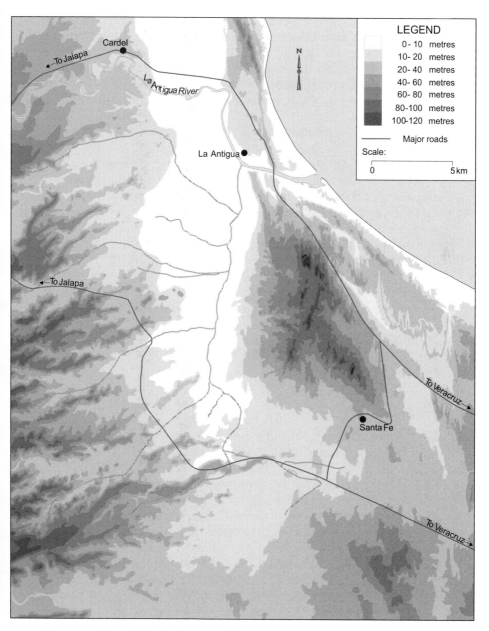

Fig. 2.2. Hills and lowlands west of Veracruz, with preintervention streams.

The central geographic expression of the shallow saucer, for our purposes, is the distribution of the remains of Prehispanic planting platforms and canals (figure 2.3). They occupy zone *b* of the wetland model presented toward the end of the last chapter. The dunes to the east and the outliers of older sediments to the west comprise what is designated as zone *a;* Laguna Catarina, varying in size and shape from year to year, is zone *c.* The three zones thus obtain more or less concentrically all around the depression, making it unequivocally an ecological and cultural basin.

The very difficulty of seeing the Basin and its absence from important historical materials, over against the concentricity demonstrable under certain criteria, have promised intriguing perspectives from the beginning and the emergence of an unappreciated instructive entity.

We deduce the Basin here as it was around A.D. 500, the time of a substantial agricultural incursion into the wetlands. This lays down a kind of datum plane. There was occupance and activity in Central Veracruz well before this, of course. And it will be necessary at some points to press the discussion back into the antecedents of the patterning. It is possible, for instance, to hypothesize fairly plausibly on how the intensification that it indicates took place. However, the gist of the discussion will be forward, through a succession of approaches to the Basin, into our time.

The Nature of the Basin

The characterization is perhaps best begun more or less as geographers have long tended to do it, with a consideration of a sequence of physical characteristics. This also creates a fund from which to draw at intervals throughout the book.

Fortunately, there is excellent raw material for the definition and characterization of the San Juan Basin. Good vertical air photo coverage at 1:60,000 and various series of topographic map coverage were freely available from official sources (e.g., INEGI 1981). Some determined sleuthing and the intervention of a friend brought to light much better air-photo coverage at 1:8000 and topographic maps that had been made from that same imagery, at scales of 1:5000 and 1:20,000, with 1-metre contour intervals (CMA 1973). Then there is the archive of my own oblique air photographs taken over the Basin during the course of more than a decade. A great deal of useful cartographic and other material may also be gleaned out of the literature on the region too, of course. More important

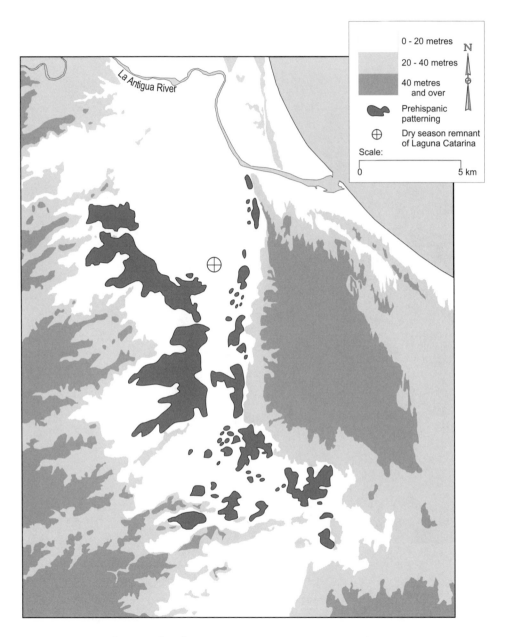

Fig. 2.3. Distribution of Prehispanic patterning in its topographic context.

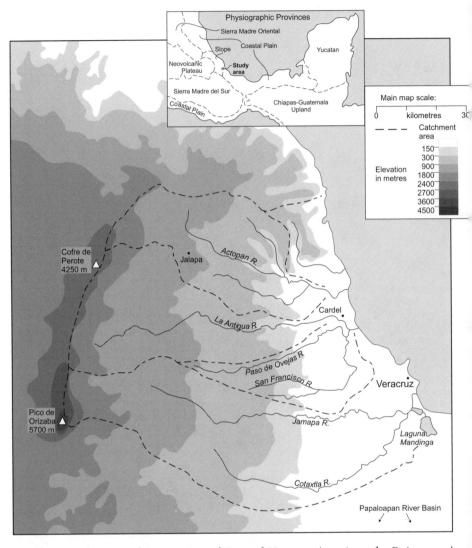

Fig. 2.4. Physiographic overview of Central Veracruz (provinces by Raisz 1959).

than anything, however, are the impressions and the pages of notes gained during deliberate and repeated ground reconnaissance.

A Sloping Piedmont and an Alluvial Plain (figures 2.4 and 2.5)

The Central Veracruzan terrain slopes eastward from the "uplifted" mountainous region. Cenozoic extrusive and older intrusive rocks underlie this piedmont, Tertiary sediments surface it, and the whole sloping formation is dissected by deep canyons. Its eastern extremity is hill land that runs out, finger-like, into Quaternary sediments, such as those of the San Juan Basin. A belt of north-south trending dunes as high as 150 metres fringe the Basin to the east. The intergraded levees of the La Antigua and Paso de Ovejas rivers fringe it on the north.

The Basin may be thought of as an embayment into which marine waters deposited fine sand (Hebda, Siemens, and Robertson 1991). Progradation of the La Antigua River delta on the north and the development of coastal dunes on the east closed the embayment. The northern portions of the Basin received more of the river-borne sediments, and much of what they received was coarse. The southern portions received much less sediment, and the deposition was

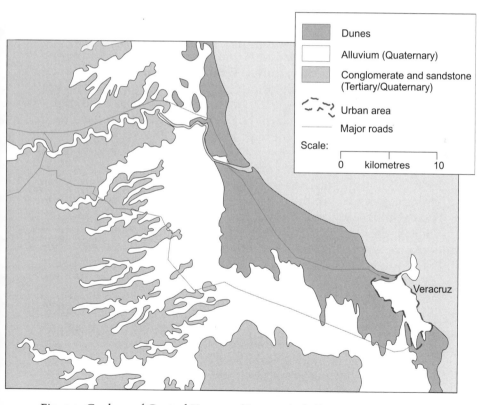

Fig. 2.5. Geology of Central Veracruz (Secretaría de Programación y Presupuesto 1984).

in the form of fine sediments: silt and clay, coming mostly during the annual floods of the La Antigua and Paso de Ovejas Rivers. Quiet backwaters were almost starved of such sediments and became sites of marl deposition. In many places marshes or swamps developed, in which organic detritus accumulated. The emergent surface eventually became agriculturally attractive to the inhabitants of the surroundings.

If one considers the fragmentation of the complexes of remains shown in figure 2.3, it is clear that a larger-scaled physical environmental differentiation is indicated. The complexes are in places set off from the edges of the Basin by colluviation, the erosion and deposition of materials from terra firma to wetlands; in others the complexes are subdivided by alluviation, mainly the deposition of levees on major and minor streams, and often interrupted by outcrops of underlying sediments related to the formations off to the west or, indeed, outliers of the dunes to the east.

Vegetational Indications

"Natural vegetation" is a fiction, but it is still useful for clarifying context, even in much-perturbed Central Veracruz. The characteristic vegetation of the sloping hill land to the west of the Basin has been designated as *selva baja caducifolia:* low deciduous tropical forest (Gómez-Pompa 1973). It is dominated by trees less than fifteen metres high and contains many xerophytes. The dunes on the other side of the basin are covered by closed scrubby thickets with a good deal of thorny vegetation (*matorrales*) on those surfaces that have been stabilized and discontinuous mostly herbaceous communities where the sand is still moving. The natural vegetation that once covered most of the Basin itself, outside of the wetlands proper, may be called *selva mediana subcaducifolia,* that is semi-evergreen tropical forest of medium height. This forest type has a partly deciduous upper story of trees that reach twenty-five to thirty metres in height.

Widespread clearing has reduced all this to remnants or successional communities. These may be seen on the "islands" in the bottomlands that often show the remains of Prehispanic settlement. They frequently make up a curious band that is characteristic of the upper limit of inundation against cultivated land or pasture on terra firma. I have only a very tentative explanation. Agricultural clearing often stops just short of the hydrophytes, perhaps because the residual moisture along the wetland margins and the associated luxuriance inhibit both the agriculturalist and his fires. The exposure of a forest edge to sunlight greatly stimulates growth in the lower stories. It becomes a reservoir of plant growth, a band of trees knit together by undergrowth, making it difficult to enter the wetlands, should one want to—a kind of ecotone of an ecotone.

Lush riparian communities remain along the La Antigua, the Paso de Ovejas, and the lower course of the San Juan, with evergreen species such as figs (*Ficus*) and willows (*Salix*), as well as shrubby and herbaceous strata. Semi-aquatic communities prevail over the true bottomland, the terrain that emerges, dries, and cracks late in the dry season. Aquatic communities surround the ponds or lakes that persist in the very lowest places. Above the lowest bottomland and below the tree growth on the "bands," the levee tops, and the "islands" there is the zone *b* of our wetland model.

The gentle swales that remain of the ancient canals in this zone are occupied by several associations of hydrophytes, which have been carefully distinguished by Gutiérrez and Zolá (1987). These have been very important in the decipherment of the Basin, since the contrast between the tones of this vegetation and that on adjacent remains of planting platforms, particularly during the dry season, constitutes our primary "inscription."

Patterning here in the Basin and elsewhere in the lowlands is often graced by palms (Frontispiece), including the *coyol* (*Acrocomia mexicana* or *Scheelea liebmannii*), which yields edible nuts, and the stately *yagua* (*Roystonea dunlapiana*). Pyrophytic and hydrophytic, these trees are usually taken to indicate long periods of human perturbation. Clearing for, and maintenance of, pastures commonly leaves a scattering of palms, which I have seen often enough in Veracruz from the 1960s onward. They may thus be associated with ranching and hence considered post-Contact, but palms are also mentioned in early colonial documents as common around native settlements in the region (Sluyter 1995: 128–129) and hence indicate earlier associations. A schematic profile of the microtopography of a wetland margin, and the associated vegetation, is presented in figure 2.6.

Aside from these intriguing remnants, indigenous and introduced grasses now dominate the Basin proper. A patchwork of crops and scrub growth on fallow extends over much of the adjacent hill land and a good deal of the dunes.

A Subhumid Enclave

The seasonal rhythm of temperature and precipitation is immediately clear from the climatographs of Jalapa, Rinconada, and the

Fig. 2.6. Sample of the vegetation over the patterning: (a) platform, surrounded by canals, surmounted by the palm *Roystonea dunlapiana,* successional remnants of *selva mediana subcaducifolia,* and herbaceous plants, especially *Leersia hexandra, Echinocloa* sp., *Cyperus* sp.; (b) canal, characterized by hydrophytes—*Pontederia sagitatta, Hymenachne amplexicaulis, Echinocloa* sp.; (c) platform surrounded by canals on three sides, typically covered with *Cyperus articulatus* and *Leersia hexandra;* (d) successional remnants of *selva mediana subcaducifolia,* making up the typical "band" around the upper limits of flooding; (e) various induced or introduced grasses and remnants of *selva baja caducifolia* (Siemens et al. 1988).

port of Veracruz; these also provide a rough climatic profile of the lowlands (figure 2.7). The temperatures are moderate in Jalapa—in overall level and seasonal variation. Levels and seasonal variations increase downslope and then decrease again on the coast. Precipitation data tend in the opposite direction. Jalapa has always had a paradisiacal reputation. Moving downslope, one comes into what is often regarded as a semi-arid or at least subhumid enclave, which extends over much of the San Juan Basin. Isohyets show this clearly (figure 2.8). The seasonal metronome beats strongly here; almost 80 percent of the approximately 1250 millimeters of its average annual precipitation falls from June to September (Gutiérrez and Zolá 1987: 23). Veracruz itself is still definitely hot and humid, but its temperatures are eased by seas breezes; precipitation figures climb back out of the enclave just along the coast.

The reasons for the enclave are to be sought in topographic eminences to the north and southeast. The Basin and the hill land to the west lie within the winter rain shadow of the mountains to the north. When continental moisture-laden air masses sweep southward at intervals (*nortes*), winter or dry-season precipitation is mostly prevented. The area also lies in the summer rain shadow of the Tuxtlas massif to the southeast, which somewhat impedes the summer's moisture-laden easterlies and lessens the wet-season precipitation (García 1976).

In the Basin a somewhat lower precipitation and particularly strong seasonal contrasts are superimposed on both hill land and rich wetland resources. Seasonally and areally interdigitated cropping thus seems suggested by the landscape itself. The natural conditions were certainly there to facilitate intensification once that was necessary—for whatever reason.

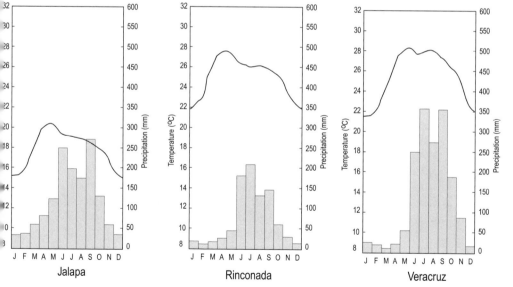

Fig. 2.7. Representative climatic graphs (Soto y García 1989).

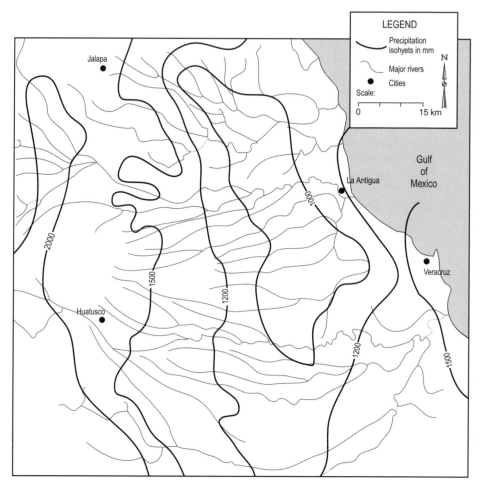

Fig. 2.8. Isohyets over Central Veracruz (García 1976).

The critical yearly downbeat is the onset of the rains. By late May local inhabitants pray for rain; the investigators, just released from their North American university year's teaching obligations, pray or at least hope that it will hold off just a bit longer. When it does come conditions for fieldwork change within a matter of hours. Pits still in process flood very quickly, and the intriguing world of the bottomlands that could be crossed in any direction becomes impassable. The rainy season is interrupted by a slackening in August; this is the *canícula*, which may imperil maize maturation on the rain-fed hill land of the surroundings (Sluyter 1995: 194–195).

Various climatic hazards have always overshadowed this region. There is the problem of drought, particularly the delay of the onset of the rains. During the dry season, or the northern hemisphere's winter, cold continental air masses move southward at intervals, gathering moisture over the Gulf, bringing wind and rain. These *nortes* can seriously damage tree and field crops; they come during the growing season on the *orilla*. Woven cane fences are occasionally erected around particularly vulnerable field crops. *Casuarina* trees line the horizon in various places, oriented against the *nortes* but also against the hurricanes. These sometimes sweep in over the region from the east, during the fall, devastating tree and field crops. And, of course, there is the occasional catastrophic flood.

Ambiguous Streams and a Fluctuating Lake

The overall hydrography of this landscape is fairly obvious; clarifying the specifics has been difficult. The hydraulic works of various scales over the past century have obliterated or at least obscured much of the anterior hydrography. Straightened riverbeds have been superimposed on sinuous earlier curves, in aid of more rapid and efficient outflow. Earlier hydrography, where it does show through, is often confusing, since there seem to have been shifts in some of the channels, particularly that of the San Juan River itself. Moreover, each series of vertical or oblique air photographs represents a hydrographic vignette. The 1:8000 vertical coverage (CMA 1973), for example, was taken at a time of high water levels; the images are sharp and clear but the hydrography has been vastly confused by floodwaters.

The La Antigua River has been monitored for considerable periods (figure 2.9), like the other major, through-flowing streams of the lowlands, but the San Juan River has not. It is thus impossible to obtain accurate longitudinal data on the behavior of even the main stream of the Basin. One observes what one can in the excavations and their surroundings and then asks local residents about the normal behavior of the water, as well as the occasional extremes and the vagaries of flow. Although such information can be diagnostic, it remains impressionistic.

Topographic maps can be hydrographically treacherous too. The SRH 1:5000 coverage with contour intervals of one metre (CMA 1973) is marvelously instructive in many ways, but confusing also, as when it slavishly follows contours at the junctions of man-made canals and bursts out in crows feet. The makers of the smaller-scaled topographic maps in the various national series have clearly

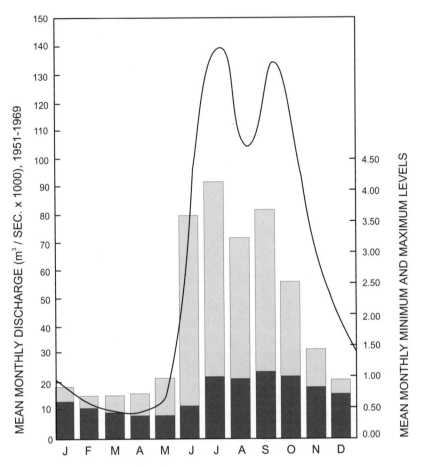

Fig. 2.9. Average annual fluctuation in the level of the La Antigua River at Cardel (SRH 1971b).

not known quite what to do with the standing water or the streams in the Basin; they have apparently worked with air photography taken at different seasons when preparing the different map sheets. Not only the features but also their names are often inconsistent.

Three perennial upland streams enter the San Juan River and its headward extension, the San Francisco River, from the west-southwest: the Paso de Ovejas River, the Arroyo San Juan, and the Arroyo Paso Real (*arroyo* is usually taken to mean intermittent, but not always!). These streams drain elongated basins that rise on the slopes of the Sierra Madre Oriental and are lodged for much of their length in canyons that striate the piedmont. They are joined by many smaller and actually intermittent streams. The dunes to the east are drained by short, intermittent streams into lagoons among the dunes or westward into the San Juan River. The San Juan, in turn, empties—if that is not too forceful a term for this sluggish confluence—into the La Antigua.

A map of the Basin was drawn in 1947 (SRH 1947) by an engineer in preparation for a major intervention, of which a good deal more will be said in chapter 7. This allows us to visualize the hydrography before it was massively rearranged (figure 2.10) and is more helpful in the physical characterization of the Basin than a current map of its hydrography.

The lowest parts of the Basin, before or after, are occupied by Laguna Catarina. It is an unassuming series of discontinuous, shallow water bodies off to the west of the San Juan River at the height of the dry season and then swells up into a lake that occupies a good deal of the northern part of the Basin floor during the wet season. The SRH engineer saw it during his fieldwork in June and July of 1947, quite likely after it had already been enlarged somewhat from its minimum by a rainstorm or two.

Variations in stream levels and flooding in this system are the result of the pattern of precipitation in the mountainous region to the west as well as over the Basin itself, the first affecting the La Antigua more than the San Juan, and vice versa. However, an analysis of inflow and outflow, which might have aided the interpretation of the functioning of the planting platforms and canals considerably, is not feasible from the data at hand. The fluctuation of flow and the changes in level of the La Antigua at Cardel are apparent on the accompanying graph (figure 2.9)—this is one of the few sets of measurements that are available for the interpretation of the hydrology of our region. The level of water as a key indicator is well understood by the inhabitants, and they are often able to recollect it in general terms for major and minor streams. A fundamental contrast soon becomes apparent: the through-flowing streams normally vary well over five metres; the San Juan varies only one or two metres.

The normal yearly rise of water levels in the Basin brings "benign" flooding to various depths. In the past, floodwaters persisted for months; today even the higher levels last for only a few days because the recently installed drainage systems remove the water rapidly. Extraordinarily high levels in the La Antigua and Paso de Ovejas rivers have occasionally led to catastrophic flooding along their banks, which is another matter.

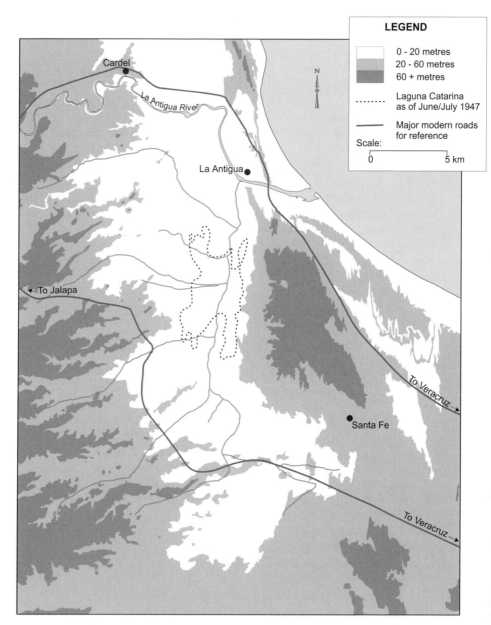

Fig. 2.10. Hydrography before the intervention by the Secretaría de
Recursos Hidráulicos (SRH 1947).

Flooding

It is useful to consider this phenomenon more closely, since it sets parameters, varies from year to year, and is rich in its effects. Approaching it frontally as a research topic yields some strong impressions. The massive literature on flood forecasting and prevention, the many proud engineering solutions, the damage assessments, and a great deal more gives the word *flooding* an ominous tone; it primarily represents a catastrophic event—with ample reason, of course. Flooding is not easily seen as benign, but that is the perspective that needs to be explored here.

It was very gratifying, therefore, to come on the work of Welcomme, in particular his *Fisheries Ecology of Floodplain Rivers* (1979), and to pursue some of the literature he makes accessible. It becomes an elaboration on things already known or suspected. Welcomme's data are predominantly African, with some Asian allusions, but this does not distract; it adds, especially given the author's many very plausible verbal and graphic generalizations. His "fisheries ecology" is a premodernization ecology, setting aside the massive recent urbanization and expansion of agriculture in tropical floodplains, and opens up a fish's eye view on the activities that interest us. A consideration of the animals that have adapted to the seasonal hydrological regime of tropical rivers and an understanding of the various opportunities and constraints for them on floodplains leads to an enlarged appreciation of the gist and detail of the Prehispanic human activities in the San Juan Basin.

Observations

One cannot normally drop everything and fly several thousand miles to watch a flood, as interesting as that might be, but one can ask about the phenomenon whenever one arrives. For example, the people of the town of La Antigua, at the junction of the San Juan and La Antigua rivers, explain that when the San Juan Basin is well and truly flooded by the reversal of the San Juan River and overspill from both the Paso de Ovejas and La Antigua rivers, continued high water in the channel of La Antigua may act as a dam on the mouth of the San Juan River. A rapid drop in the level of the former, say with the abrupt cessation of rains in the mountains, allows the water of the Basin to rush out and over the channel of the La Antigua, flooding the town. Benign flooding has become destructive.

One can also happen in on a flood, as we have done on several occasions. In early July of 1991 a high but not yet catastrophic flood took place in the San Juan Basin. We came by a few weeks after the water had gone down. Informants recounted that water levels had risen within a matter of hours, as happens often. The floodwaters came down again in days, as a consequence of the various private and governmental drainage works now in place. This had a curious concomitant, an echo of the old benefits of benign flooding. Freshwater shrimp (*Macrobrachium acanturus* and *Macrobrachium carcinus*) entered the wetlands with the floodwaters and dispersed (Olguín, personal communication 1991). The rapid descent left them stranded. Local inhabitants gathered as many as they could out of the pastures—a fine "harvest." Later an unpleasant smell spread through the wetlands as the remainder rotted. Evidently fish, as well, often migrate into the wetlands during such floods and then frantically seek out the deeper places as the waters recede. There is excellent fishing then for some days.

The air obliques we took in mid-July show that silt was deposited widely within the Basin, leaving a line against the sides of low hills. We were able to deduce a flood that had come to about five metres above mean sea level. The corrals and milking sheds on top of these "islands" had been left dry.

In June of 1993 another high but not yet destructive flood occurred. Approaching the Basin by air soon after the crest, we found the bottomlands one vast lake (figure 2.1). Laguna Catarina, which is a shrunken affair of perhaps a kilometre or two in diameter during a normal dry season, had reasserted itself. Our oblique air photographs allowed us to approximate the margins of open water and to relate the lake to the distribution of patterning (figure 2.11). This has stimulated various reflections. Could one imagine the wetland agricultural system represented by the patterning to operate under hydrological conditions such as these?

Some specific circumstances have to be noted. This was the flood of one year; we have only ethnohydrographic evidence to indicate that it was a relatively high flood. We had caught this one just after its crest, and there are all those many modern drainage systems built precisely to draw down floodwaters around the margin of the wetland rapidly. So the lake will have been larger shortly before the photograph in figure 2.1 was taken. In order to distinguish between flooded and nonflooded land cartographically, open water was enclosed. However, this boundary might well leave considerable areas of saturated land outside. It would not be feasible

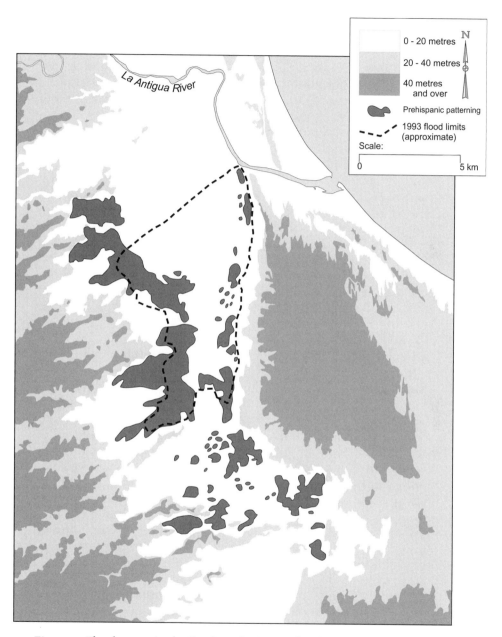

Fig. 2.11. Floodwaters in the San Juan Basin in July of 1993 superimposed on the distribution of patterning.

to work this land while saturated; plants would not be able to find the aeration that they need, but at the same time the benefits of flooding would accrue as well. Therefore, the land is likely to have been "flooded" beyond our boundary of the lake (figure 2.11), not so much eastward toward the relatively abrupt boundary with the dunes, but westward into the very gently upward sloping terrain of the valleys between the Tertiary upland extensions, thus encompassing a good deal more of the patterning.

Also, we have often been told of, and have actually seen, local ponding, which is elusive from the air unless the sun happens to flash up suddenly from a pasture. Usually in the wet season a perched lagoon extends over a part of the Nevería complex of patterning, which will be discussed in the following chapter. Road building and canalization have altered the relevant areas to the southwest of the complex, but it is likely that a larger lagoon was once retained seasonally over a part of the complex by the remains of older sediments trending in fingers up from the southwest. Similar conditions may be postulated for the other patterned areas to the south of the limits of the lake as observed in 1991, accounting for more of the patterning still.

Various factors complicate any attempt to clarify relationships, particularly the link between flood levels to sea levels from the time of the construction of the fields to the present. It seems we are within the range of "noise" on any creditable, necessarily longer-range plotting of sea level changes. It does not seem feasible, for the period under consideration, to distinguish the effects of slight secular climatic variations, or simply climatic vagaries, as well as intervening alluviation, possible tectonic shifts, and slight sea-level change. All or any of these factors may have been counteracting or reinforcing one another.

In any case, such gradual environmental changes may well have been counteracted by the agriculturalists' adjustments in the wetland fields system. It could be altered by progressive build-up or build-down. And it could be used variously throughout the vertical range of flooding or saturation: beginning a cropping sequence further upslope or ending it further downslope.

It is possible then, by and large, to identify patterning with land subject to inundation, under conditions more or less as they are today. From the beginning of our whole long investigation of "raised fields," while the model was taking shape, this was a basic postulation. Here, and in the evidence to be presented later for sedimentation mingled with that for the agricultural use of the fields, were strong indications that the relationship was not unreasonable.

Synthesized Normal Course of Benign Flooding

Tropical lowland floodplains and their immediate surroundings pulse with the seasons. The hydrological variations may be expressed in average maximum and minimum water levels. There is systematic data on main through-flowing streams (e.g., figure 2.9); elsewhere such levels are easily pointed out by informants against something vertical and relatively permanent.

A normal flood is phased. The first rains over a basin quickly saturate the portions of wetland soils that have emerged during the dry season; ponds develop and an outward flow begins in the seasonal streams. Once the effects of rains in the adjacent mountains have worked their way down the through-flowing streams, water begins to flow in the opposite direction, over the lower sections of the levees on to the floodplain. This is mostly a slow creeping flow, during which the sediments in suspension are precipitated. These are the waters that allow fish in the streams access to habitats formed seasonally on the floodplain beyond the levee. Water levels will drop again with evaporation, outflow from the wetlands by means of the seasonal streams, and seepage from the wetlands into the channel of the adjacent through-flowing stream. The key variable for the plants and animals of the wetlands and for the people who attempt to exploit this environment is the length of time floodwater covers any given part of a wetland. From early human incursions into wetlands to the present time it seems to have been important both to accelerate the beginning of the water's recession after the end of the rains by means of canalization and to retard the outflow in the dry season, both of which are advantageous to cropping.

There are extremes, particularly the extreme highs that may be several times the average highs. They may be deduced from the hydrological data for the through-flowing streams and, of course, are vivid in the memories of the informants. This is when the benign becomes catastrophic.

Less dramatic variations may be expected from year to year. A "poor" flood reaches only a low maximum level and fails to cover wetlands entirely, thus minimizing fish habitats. It drains quickly, leaving only a skimpy deposit of silt, and remains generally inefficient in its various "fallowing" effects. A succession of such floods is particularly distressing. A "good" flood is the opposite in all of these respects. These are the conditions one must gamble on or learn to use. All of it is surrounded by folklore and endless talk; it will once have been the subject of ritual.

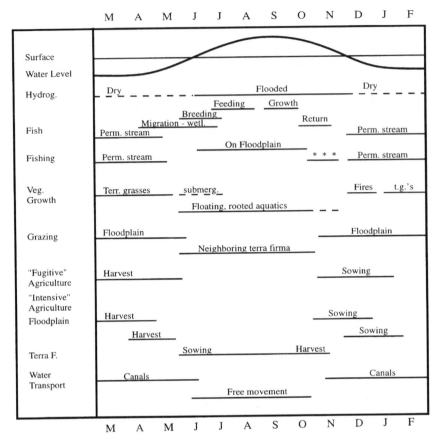

Fig. 2.12. Seasonality in and around a tropical lowland floodplain.

The yearly rise and fall of the water level and much of what follows in an essentially unmodernized floodplain during the course of a year may be synchronized, as in figure 2.12, which relates vegetational changes, the movements of the fish, the opportunities for fishing, and the conditions for human movement over the floodplain, as well as the various phases of agriculture and ranching.

Flooding is physically prejudicial to any earthen structures, even where the water only creeps in and rises slowly. Platform and canal sides will have been vulnerable to slumping. Redigging and reinforcement are likely to have been the most laborious aspects of the system—acceptable as long as the yields were needed and social conditions permitted. Left untended, the planting platforms and canals would soon be smoothed into gentle waves and the remains of the platforms colonized by trees.

However, there are many positive effects of flooding. In the first instance, benign flooding recharges soil moisture, maintaining the greenery, the swamps, and the lagoons that will provide oases when all else has turned to grays and browns in the dry season. The movements of fish further articulate these general effects. Many need to migrate considerable distances between their dry-season and wet-season habitats—out of the main stream channel into the various other water bodies of the floodplain and back again, if they can. The main factor influencing their distribution within the aquatic system at any given time is the distribution of dissolved oxygen. They tend to concentrate in those pools or lagoons on the floodplain where there is sufficient oxygen, as though these were "sumps." This concentration provides excellent opportunities for natural predators, particularly birds. Waterfowl are very common in wetlands anyway; their life-cycle is closely linked to the floods. Those that take fish can do very well in the dry season, as can the people of the surroundings. Wetland resources for fishing and hunting must always be considered as a significant part of traditional subsistence systems in the humid lowlands. Even where modernization has diminished the significance of working on the land, people are likely to go out into the wetlands occasionally to hunt and to fish.

The entry, nurture, and concentration of fish have been managed in traditional ways in various places. Welcomme describes an interesting example from West Africa (1979: 196–201, 240–241) and cites comparable practices from elsewhere (Chevey and Le Poulain 1940; Hurault 1965). Channels are dug through levees to facilitate the passage of the fish from the main stream onto the floodplain. Here long, thin "drain-in" ponds are excavated, looking very similar to canals cut for other purposes but serving, as do the natural "sumps," for the concentration of fish. Vegetation is allowed to cover much of the surface, under which the fish thrive. Inhabitants progressively block the ponds with bamboo barriers and remove the vegetation until the fish are enclosed within a small space, ready to be harvested.

The morphology of the West African examples of incisions into a floodplain, when seen from the air (Welcomme 1979: 199–200, 240–241) has been helpful in the interpretation of the vestiges of human incursions into the wetlands of lowland Mesoamerica. Early in our investigations of the Prehispanic remains of this landscape, we noted extensive webs of linear incisions on the flood-plain of the Candelaria River, which did not make sense as elements in either an agricultural or a transportational system, but rather as the remains of an impressive fishery. The morphology of the African example is even closer to the remains of incisions found on the floodplain of the Río Hondo in Northern Belize and others noticed on various floodplains of Northern Veracruz. Such canals have not been found in the San Juan Basin, which does not mean that seasonal fishing was not significant. The fish could well have been directed into the webs of canals amongst the planting platforms and harvested there, too, in due time with drives and nets.

The water that enters the floodplain by the reversal of the flow of seasonal streams and overspill from the channel of the through-flowing streams is charged with suspended sediments (Petts and Foster 1985: 115–119). These include mineral particles corresponding to the geology of the catchment basin, as well as organic debris derived from terrestrial and aquatic sources. The concentration of these sediments varies with the seasons, the peaks related to flood-water peaks but often lagging somewhat. The larger particles drop out soon after the water leaves the main channel; smaller particles, making up "silt," are carried out onto the floodplain, where they settle out.

Silt has long enriched floodplains important to human history; the seasonal gift of the Nile is common knowledge, as is the vitiation of that largesse with the building of the Aswan High Dam. Worked in, together with the detritus of the vegetation killed by flooding, silt can substantially enrich alluvial soils. It will have been valuable on the surfaces of the ancient planting platforms and in the canals too, where it would combine with organic debris to form muck that could be scooped onto the platforms as a very effective fertilizer. This is done to great advantage in the chinampas south of Mexico City to this day.

The advancing waters enrich in another, chemical sense (Welcomme 1979: 47–49). Conductivity is a measure of the ions in a body of water and an approximation of its chemical richness. In the vanguard of the advancing floodwater, the contact of the water with soil produces a local increase in conductivity, which makes these moving fronts biotically highly productive.

The vegetation of a floodplain is closely adapted to the rise and fall of water levels and hence strongly zoned vertically (Welcomme 1979: 58–67). It is useful to follow this differentiation here, since it helps in the understanding of the effect of flooding and also the manipulation of various ecological propositions already broached.

At one extreme is the submerged vegetation of the permanently flooded zone with open water, then the rooted or floating emergent vegetation of the somewhat shallower regions—both in the zone *c* of our simplistic model (figure 1.9). Then there are the zones that are regularly flooded seasonally, with rooted and floating emergent vegetation, as well as those areas occasionally flooded (that is, between mean and highest seasonal flood levels), with their mostly rooted vegetation. This is our zone *b*, the margin of the wetland, the context of the vestiges of planting platforms and canals. At the upper extreme are the areas that are not flooded, except perhaps during catastrophically high levels, but whose water table is influenced by flooding. These are the levee tops and lower slopes of terra firma—toward the lower end of our zone *a*.

Flooding drastically reduces gas exchange between soil and air; microorganisms soon consume almost all the oxygen in the soil and in the water just above it, too, if the water remains stagnant. The results are marvelously complex but may be simplified for our purposes: plants not resistant to flooding are stunted or killed altogether (Kozlowski 1984: 3–5). The soil is likely to be quickly reaerated after the descent of the floodwaters. The length of time of submergence is a key variable for natural and introduced plants on the floodplain. It is obviously important to use cultivars with convenient growing times.

The flooding of an agricultural surface within the floodplain soon kills the cultivars and introduced "weeds" adapted only to unsaturated soils. Rhizomatous "weeds" easily survive flooding; they thrive after the water goes down and have to be combated by other means. Large portions of the floodplains of Mesoamerica are naturally covered by grasses that propagate by means of underground stems and that are actually stimulated by flooding. Introduced grasses largely propagate and are stimulated in the same fashion. In the dry season, the grass growth, now dry and hard, is normally burned in order to allow a new greening. Certain trees also are adapted to flooding; the stately *yagua* palm (*Roystonea dunlapiana*) found in magnificent stands in the wetlands of the San Juan Basin is both flood and fire resistant.

And flooding has a further effect. Waterlogging favors microorganisms that are antagonistic to plant pathogens. Infection and disease development can thus be reduced by inundation (Kozlowski 1984: 255–256). Taken altogether, this reduction of disease, the killing of unadapted plants, and the precipitation of suspended sediments would seem to constitute an excellent fallow. This is not just benign but beneficent.

There is an iconographic indication that these various positive effects were already appreciated around the time that the planting platforms and canals of Central Veracruz were in operation. Professor Alfonso Medellín of Xalapa, who was highly supportive of our early work in the wetlands of Central Veracruz, interpreted a relevant fragment of a mural found at the Late Classic (A.D. 600–900) site of Las Higueras, some 120 kilometres northwest of the port of Veracruz (Medellín 1979). In this fragment a god and goddess of inundation are represented; between them is a maize plant with ears emerging. The god is pouring out a jug of water over the earth, like a blessing, and in the current there is a fish. Flooding seems represented as the very means of productivity.

An Aerial Sampling

A most absorbing aspect of the study of the Basin has been the scrutiny of the 1:8000 vertical air photographs. However, these are best handled like a deck of cards, fanned, scooped up, reshuffled, and then dealt under a stereoscope. One must be able to descend into them. It would be difficult here to achieve images with sufficient resolution and in manageable stereo pairs to allow the reader anything like such a descent. Oblique air photographs are another matter. Their scales are usually larger, they can be more easily manipulated in the printing process to heighten contrast, and the content is directly accessible to the viewer.

A suite of oblique air photographs and interpretive maps are presented in order to allow the reader to dip into the richness and complexity of this landscape (figure 2.13). It will be a graphic restatement of the context of the patterning and a beginning of the discussion of its morphology. Three views show substantial complexes of agricultural remains and their immediate surroundings; they are accompanied by tracings of the mid-lines of canals within the complexes of planting platforms. A topographic map shows one of the substantial settlement sites on the alluvium surrounding the Basin. One further view is presented from adjacent hill land, also accompanied by a tracing, showing the remains of an agriculture that is apparently related to that practiced in the wetlands.

Canal midlines seemed the best way of grasping the morphology of the agricultural system. The remains of planting platforms are eroded remnants, so any attempt to show the system by circumscribing the remnants would underrepresent the original planting surfaces and overrepresent the canals; it would also obscure

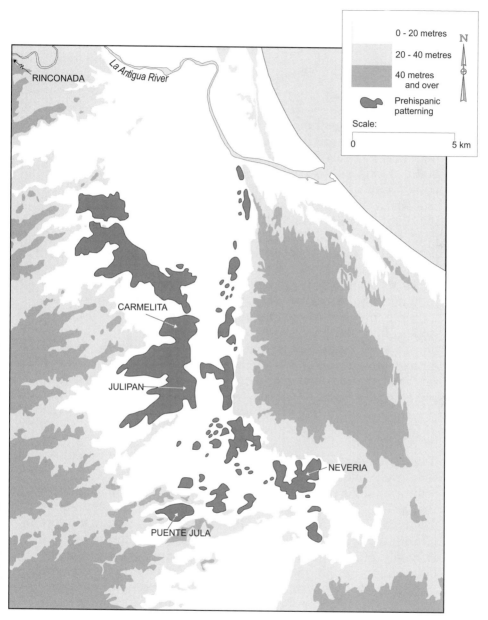

Fig. 2.13. Locational map for figures 2.14–2.21.

morphological trends in the overall complex. The tracings are syntheses of patterning visible on vertical air photographs and on various runs of oblique black-and-white, as well as color, photographs.

Tulipan

Substantial and striking, the complex of remains known as Tulipan was named for the ranch that has its headquarters nearby (figure 2.14 and 2.15). It is found within one of the "optimal" wetland contexts outlined toward the end of chapter 1. In photographs, the platforms are mostly light and the canals dark. They are not random in their orientation; they vary in form from the edges of the complex toward the center. The patterning is interrupted by a grove of trees on an "island" with evidence of Prehispanic occupance. A fairly clear boundary between patterned wetland and terra firma is apparent in the top left corner of figure 2.14.

The most obtrusive present geometry of the wetlands follows from fencing and variations in the management of pasture, particularly the intensity of grazing and the differences between the varieties of grasses. Tree growth is only a residuum: the light spiky growth in wetlands indicates various species of palms; the rounded forms represent succession species of *selva mediana subcaducifolia*. The latter may appear as live fences or as clusters on "islands" and along the boundary of wetland and terra firma.

Carmelita

The patterning of the Carmelita Complex also occurs in an "optimal" occluded wetland—between the arms of higher tertiary sediments (figures 2.16, 2.17). In the figures, the higher area is to the left; the complex runs out toward Laguna Catarina on the right. The lighter and darker tones on the two sides of the *orilla* are very clear in the upper left corner of figure 2.16. Again, the orientation is hardly random. Staring at the labyrinth in the center of the oblique photograph one begins almost involuntarily to trace the possible routes canoes might take through the system. With only slightly more imagination, the dry-season crops themselves take shape. One wonders if there were trees firming up the edges of the platforms as in the chinampas south of Mexico City. If not, many, many embankments will need to have been refurbished after each flood.

Fig. 2.14. Air oblique photograph of the Tulipan complex of planting platform and canal remains. North arrow and encircled features relate figure 2.14 to 2.15.

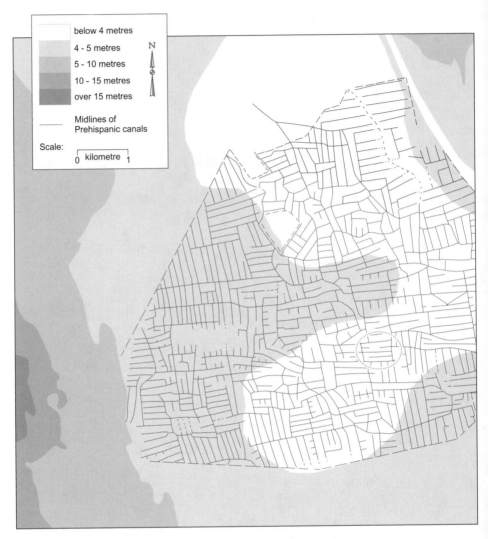

Fig. 2.15. Canal midlines for the Tulipan complex.

Historical drainage canals slice through the remains. The brush that has been allowed to grow up on the right side of the main canal simulates the vegetation that will have developed after the ancient agriculture was abandoned. That was mainly what the nineteenth-century observers noted in the Basin: a forested stretch along the road up to Jalapa. The photograph also shows how even low woody growth can completely conceal the remains.

Fig. 2.16. Air oblique photograph of the Carmelita complex. North arrow and encircled features relate figure 2.16 to 2.17.

Fig. 2.17. Canal midlines for the Carmelita complex.

Nevería

Nevería is the easily accessible ramified complex of agricultural remains to which we have given more attention than any other in the Basin (frontispiece; figures 1.1, 2.18, 2.19; Siemens et al. 1988). It lies at the upper end of the Basin, topographically speaking; i.e. between 13 and 15 metres above mean sea level, in contrast to the Tulipan, Carmelita, and other complexes to the northwest that are below 5 metres. A finger of Tertiary sediments from the southwest and dunes from the north seem to have impeded drainage northward, facilitated sedimentation behind them, and thus raised the southern extremity of the Basin.

A group of mounds is hidden among the trees at the top right-hand corner of the oblique air photograph (figure 2.18). Here is a typical ancient settlement site between two microenvironments: the sand hills, potentially productive in the wet season, and the wetland, very useful in the dry season.

Three-sided platforms fringe the complex on the southeast side and elsewhere. A quadrate pattern predominates within (figure 2.19), similar to that already illustrated for El Yagual (figure 1.1). An orientational tendency also is clear.

Pasture management superscribes the ancient patterning; various intensities of grazing are apparent, as well as some fieldwork with machines. Trenches have been cut to serve as intake pools for sprinkler irrigation systems.

Palms and successional remnants of an earlier forest dot the pastures. The convoluted boundary between the darker-toned terrain that is subject to flooding and the lighter-toned higher ground is very clear in the right bottom corner of figure 2.18.

Caño Prieto

Remains of the settlement known as Caño Prieto dapple the alluvium in the northern extremity of the Basin (figure 2.20). Mounds are the prominent features; shallow depressions are often noticeable nearby. All this is above the normal limits of inundation. There is no sign of patterning immediately around the mounds, but there *is* very extensive patterning in wetlands to the south.

It is not difficult to postulate a subsistence equation: wet-season cultivation on the alluvium, dry-season agriculture in the wetlands. If the depressions are contemporary with the structures, they are likely to have served as reservoirs during the dry seasons. Crops of various kinds, notably sugarcane and mangos, grown with the help

Fig. 2.18. Air oblique photograph of a part of the Nevería complex. North arrow and encircled features relate figure 2.18 and 2.19. Dashed arrow points to the subset El Yagual, seen in frontispiece and figure 1.1.

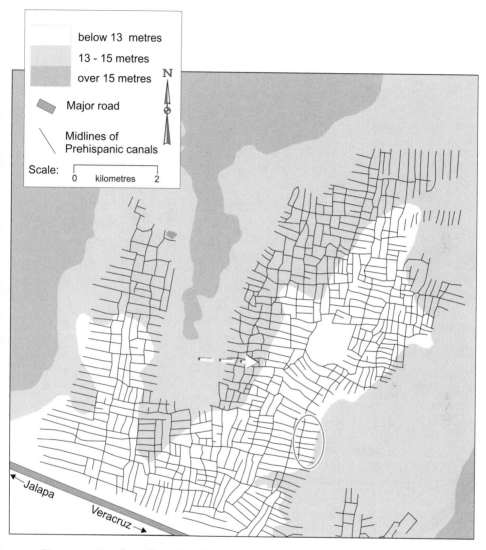

Fig. 2.19. Canal midlines for the Nevería complex.

of irrigation, now predominate over the center and the western side; ranching is dominant on the east.

Mounds with the remains of settlement and neighboring depressions were also encountered just to the south of this on the floor of Laguna Catarina. They are small tree-topped islands during high water—refuges for cattle.

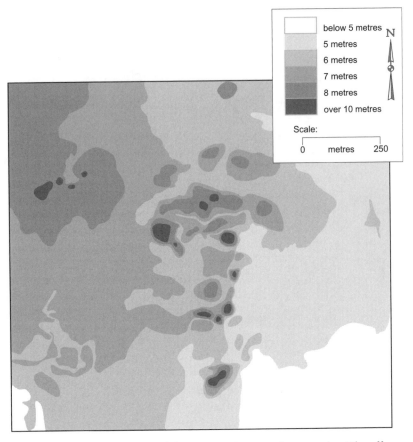

Fig. 2.20. Topographic map of the Caño Prieto settlement site. The alluvium that encloses the San Juan Basin along the La Antigua River grades down from the northwest; the Basin proper lies to the southeast.

Rinconada

Extensive series of roughly parallel and slope-transverse stone lines are to be found on the piedmont west of the Basin and just north of the town of Rinconada (figures 1.1a, 2.21, 2.22; Siemens 1989c: 197–208; Sluyter and Siemens 1992). They are the remains of Prehispanic field boundaries, best thought of, perhaps, as the retaining walls of terraces or semiterraces (West 1970). One sees them on freshly cultivated land just after the beginning of the rainy season, when windows open on the distant past. They close again with drying; the lines on cultivated and uncultivated land become invisible.

Fig. 2.21. Air oblique photograph of stone lines on the piedmont near Rinconada.

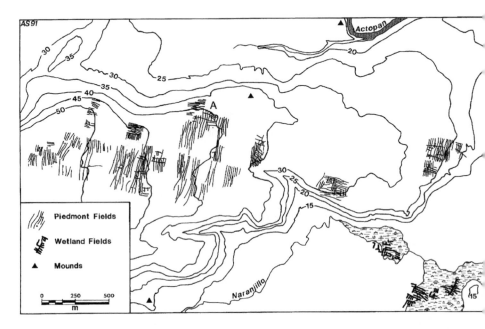

Fig. 2.22. Map of stone lines on the piedmont (after Sluyter and Siemens 1992: 152).

The search for them was stimulated by a fine nineteenth-century description of the prehistoric litter on these gentle slopes and in the canyons that slice into them (Sartorius 1961). More of that in a later chapter.

We hypothesize that the terracing represents an intensification of wet-season agriculture by means of the partial clearance of stones from what is often very stony ground and, at the same time, a conservation of moisture and the soil itself. The fields enclosed could have been used for a variety of crops; their cultivation was probably integrated with land use on wetland margins. The orientation of the lines is remarkably accordant with that of the Prehispanic canals in the wetlands. From this relationship and the ceramics in settlements closely associated with the stone lines, a similar chronology may be hypothesized as well; we do not yet have direct dates.

Summary and Links to the Model in Chapter One (figure 1.9)

A key dividing line between darker and lighter tones, between terrain that is and is not subject to inundation, recurs repeatedly.

Zone *a* lies above and zones *b* and *c* below. The "strand line" cannot be traced continuously around the Basin because of the interfingering of Tertiary and Quaternary sediments, the complication introduced by stream sinuosity, and all the tonal variations consequent on use.

Above this key line—on the "islands" and on the "promontories" of the older sediments and on the higher reaches of recent alluvium—are the remains of Prehispanic settlement. Subsequent settlement has sought out the same locations. Above the line, as well, are irrigation canals, the irrigated pastures, and commercial crop lands, plus forest remnants (*selva baja caducifolia*), succession communities, and rainfed agricultural lands in fallow and in use.

A good distance above the line, farther than one can easily grasp in one photographic frame and well beyond the Basin, are the remains of an intensification of rainfed agriculture on the sloping piedmont surface, the stone lines. They amplify the scale of the conceptualization of interdigitated cropping in lowland Central Veracruz. One may postulate it in close proximity, perhaps as practiced within communities, but also as a regional complementarity of intensified production.

Below the line is the Prehispanic patterning, scored with subsequent drainage canals of various widths cut in the context of individual or governmental projects and marked very prominently by long, straight fence lines with different degrees of clearing and grazing intensity on either side. Remnants of forest (*selva mediana perennifolia*, mostly) and succession species dot this surface, depending on how diligently the rancher has "cleaned" his pastures of woody growth. Here and there are groves of palms.

The Form of the Agricultural Remains

Visual evidence regarding the morphology of the agricultural remains has been the initial and most persistently intriguing line of evidence in the investigation of Prehispanic wetland agriculture. It seems definite and unequivocal at first, but what can be read from it remains fundamentally limited. It has often been minimized, perhaps seen as useful for introduction and illustration but not as weighty as data from the pits. In fact, of course, test results on the physical and chemical nature of the sediments, radiocarbon dates, and certainly the meager sherds one finds in this context are quite likely to be as contingent in their meaning as the forms derived from air photography. It was determined from the beginning of the investigation of Veracruzan remains to bear down hard on morphology.

"The Magic of the World"

A consideration of morphology leads fairly quickly into some old philosophical issues, including

> . . . the everlasting fascination exerted by Form. Form is both deeply material and highly spiritual. It cannot exist without a material support; it cannot be properly expressed without invoking some supra-material principle. Form poses a problem which appeals to the utmost resources of our intelligence, and it affords the means which charm our sensibility and even entice us to the verge of frenzy. Form is never trivial or indifferent; it is the magic of the world. (Dalcq 1951: 91)

There are strong constraints on the interpretation of form, particularly form that is far beyond the memory of present inhabitants and unaccompanied by documents (Hodder 1986). The central difficulty in archaeology has always been the resolution of the relationship between material culture and behavior. There is certainly no direct, universal, cross-cultural relationship. There are always shifting assortments of ideas, beliefs, and meanings interposed, which would seem to undermine attempts at scientific inquiry. There is no alternative, however, but to persist, and in particular to assess context very carefully.

The textual metaphor comes readily to mind in the interpretation of a flat wetland landscape peered at through a stereoscope. However, some qualifications must be kept in mind. Material remains are different from a written or spoken language. They seem simpler but more ambiguous, more fixed and durable, not as arbitrary as words, but more like images or representations (Hodder 1986: 176). Deductions from such a text, particularly from a suite of prehistoric agricultural remains, are likely to be small in scale and limited in resolution, but they are potentially fundamental to an understanding of society and culture. These are not incidental artifacts; they were part of an infrastructure that sustained life and that was labored over for generations. However, the understanding of their meaning is not necessarily enhanced by large-scale and scientifically sophisticated analyses. The subject and the method of study can easily go out of phase; as the measurements become more elaborate, the meaning can slip away.

A very helpful survey of the borderlands of geography and semiotics has touched on what features of the kind we are considering may be communicating:

> Material expression [results from] a wide variety of behavior based on the production, exchange, and consumption of physical goods which are employed to express, among other characteristics, identity, affiliation, affect, and intention . . . objects are long-lasting and transmit or reflect energy relatively permanently. (Foote 1985: 166–168)

The patterning in the wetlands represents venerable artifacts indeed, and we do have a glimmering of what they express regarding identity and affiliation, but most of the feelings and intentions of the builders seem beyond our means of detection. We have much better access to these aspects of the historical superscriptions.

There is a potential psychological problem here. "With the [human] capacity to imagine pattern where there is confusion goes the tendency to create cosmos where there is chaos" (Walter, in Whyte 1951: 190). In the enthusiasm that surrounded the early investigations of "raised fields" in the Maya region, particularly in Northern Belize, my collaborators and I sometimes thought we saw artificial patterning where it turned out that there was none. As one scanned forested lowland, the trees themselves would sometimes perversely line themselves up—a confusing optical effect. Flying over the bajos of the Petén, one would catch a glimpse of a line crossing one of them and hurry to photograph it; the print itself would show nothing. Radar imagery seemed to some investigators, for a time, an excellent medium with which to make cosmos out of chaos, but it did not convince, as was outlined in chapter 1.

The Canal Midline as a Graphic Device

The graphic rendition of the visual evidence has been problematic. The planting platforms and canals have been typically mapped by circumscriptions of the platform remains. This requires a large assumption on the vertical dimensions of their form. One cannot usually tell a great deal about this with or without excavation because of erosion and sedimentation after the use of the fields, as well as the various other kinds of vertical perturbation affecting all stratigraphy in these wetlands. This same argument undercuts attempts to quantify field surfaces or volumes of earth moved. Further uncertainty over the organization of work makes it very difficult indeed to deduce the labor expended in construction and maintenance. Various possible degrees of intensity in agricultural production put in doubt any calculations of carrying capacity.

An alternative was found, as already indicated, by using midlines of canals (figures 2.15, 2.17, 2.19). This method seemed quite accurate, since one can assume fairly symmetrical erosion and sedimentation in this context. The preparation of the maps was demanding; it was done with great care by Alastair Robertson. Data from the 1:8000 vertical air photography was combined with that from oblique photography on various films, taken from many angles and at different times of year. The oblique data had to be converted laboriously to a vertical plane before it was comparable. An attempt was made to do this by means of computerized photo correction, but the results were not nearly clear enough.

Such a painstaking process shows the limitations of any reading of such a landscape text. Vegetation can easily obscure, as can floodwater. One perspective reveals and another, for reasons of specific reflective conditions, covers up. Variations in the processing of the film and the printing can make a considerable difference where other factors are approximately equal. We have achieved only partial results, but these are unrivaled for scope and detail in any publication on the subject to date.

A simple superimposition of a scaled grid over the planting platforms and canals thus traceable in the Basin allowed an approximation of the area of patterning: just short of 1600 hectares. Adding 20 percent for what was not visible indicates something like 2000 hectares for the system—not the agricultural surface alone, but the system as a whole.

Representation by midlines confirms the overall gross four-way regularity in orientation already suggested in the oblique photographs (figures 2.14, 2.16, 2.18). However, the gist is subtly interrupted here and there by variations in the directions of the canals and in the ratio of platform length to width, which suggests regions. These are rather similar, as outrageous as the comparison may sound at first, to the variations in the orientations of the grid plans of Spanish American cities, which indicate discontinuities in development. It is not difficult to envisage a succession of expansions or variations in tenure in the fields.

At the same time, there are relationships between the forms of the fields and slope. This is repeatedly and very clearly apparent in the various complexes found in Northern Veracruz and occurs in the San Juan Basin too (figure 2.18). From the wetland margin against one of the many slight rises within and around the Basin toward the center one sees in turn: downslope canalization in elongated U-shapes, then quadrate canalization, and then, often, no canalization at all in the centers of the wetlands. The first of these seems to have accelerated the drying out of wetland margins at the end of the wet season and thus allowed earlier access than otherwise possible. The second retarded the outflow later in the dry season. This system represents drainage, irrigation, and even storage, a series of subtle functions achieved by simple means. A single function designation such as "drained field" thus just does not serve.

But there is something curious here. Along the "strand line," the canals that enclose the long, three-sided platforms seem frequently bent from the overall grid in order to provide more efficient drainage. There seems therefore to have been the occasional trade-off between an organizing principle and local practical necessity.

From the beginning, vestiges of enclosing or controlling features, such as substantial dams and dikes, were sought with considerable care. Sedimentation can obscure a good deal within a basin, especially in its lowest areas, but if whole complexes of platforms and canals are apparent, then it is to be expected that other related earthworks with similar masses would have left equally visible remains as well. None were found in the Basin. Admittedly, there has been much recent disturbance with the intervention of the Secretaría de Recursos Hidráulicos, but some remnants of diking should have been apparent if these features were there. This has considerable significance for the interpretation of the function of the system; it precludes the inference of a full-fledged chinampa system with its strong control of water levels.

The fabric of the canal systems typically suggests neither the confluent pattern that one would expect in a drainage system nor the distributory pattern that would indicate canal irrigation. Instead one sees baffles, which are likely to have served for the retention of sediments during flooding and perhaps the retardation of the drop of water levels in the entire wetland toward the end of the dry season—quite conceivably aided with the erection of small, temporary earthen dams within the canals. Such a canal pattern may well have facilitated fishing, making it difficult for the fish that had come into the flooded wetlands from the main stream as fry and grown up in water away from the main stream to regain that stream as levels dropped. Also, the lattice of canals would provide convenient proximity to surface water for scoop irrigation during the dry season.

Orientations

At an early stage in the investigation of wetland agriculture in Central Veracruz, the orientation of the remains in a number of wetlands, including those of the Basin, led to a cosmographic inference (Siemens 1983b). This subject often raises eyebrows; it seems suspect, insubstantial. I have heard colleagues who have entered into archaeoastronomy much further than I have describe the resistance they encountered and have felt it myself repeatedly. In the wetlands of Central Veracruz the evidence was irresistible.

Orientations had been noticed in the monumental architecture of key Mesoamerican sites for many years, but this tracing of regularities in the arrangement of ancient agricultural surfaces seemed to be a new departure (Religion and Agriculture in Mesoamerica 1983). Cardinal directions were apparent but were offset clockwise. This was tested on vertical air photography of a scale of 1:60,000, which was scanned with a magnifying stereoscope. Wherever trends were noticeable on one axis or two within substantial complexes of planting platforms and canals, they were noted on a clear plastic overlay with lines drawn freehand, rapidly and without much reflection. Too deliberate a procedure would have evaporated this evidence; the gist would have been lost in the particularities of the ancient and the superimposed features. The results were then optically reduced in scale, transferred to 1:100,000 topographic coverage, measured, and graphed by categories (figure 2.23). Of 33 vertical regularities noted, 27 deviated clockwise from north and south between 5 and 25; 15 of those, between 10 and 20. Of the 29 transverse regularities, 24 deviated clockwise from, and east and west between, 5 and 25; 15 of those, between 10 and 20. A minor vertical and transverse clustering between 5 and 10 anticlockwise of our cardinal directions also was noticeable.

Such a method of interpretation seems roughly commensurate with the way in which an orientation is likely to have been inscribed in the first place. It could be only crudely ascertained and expressed with the technology of the time. Also, we are dealing here with agricultural surfaces and not buildings with walls of masonry. Accuracy would have to be less and what might have been there at one time could easily have been distorted this way or that in maintenance. Moreover, the directions would likely have been taken over a considerable period of time by various social units as the whole system was put in place progressively within and between wetlands.

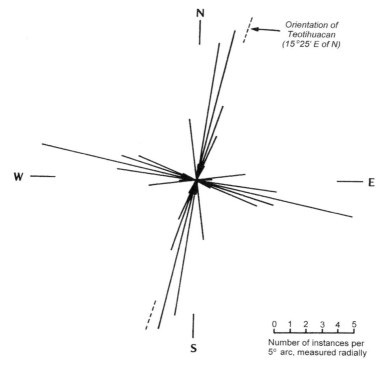

Fig. 2.23. Canal midline orientations in the wetlands of Central Veracruz (Siemens 1983b).

A more formal method was tried: a statistical analysis of orientations at randomly chosen points within complexes (Hammond and McCullagh 1974: 114). It confirmed more than it denied but glossed over a good deal and was most cumbersome. The initial, simpler, and rather impressionistic approach had been more incisive.

The cardinal directions seemed fairly clearly to be an allusion to a widespread ancient Mesoamerican view of the world as divided into four quarters, each associated with certain aspects of the world of men and of the gods (Carlson 1981: 146–154). It was hypothesized that the main, clockwise deviation from our own cardinal points indicated an organizing principle related to the sacred direction of Teotihuacán (15 degrees, 25 minutes clockwise) and that it signified the impact of the power and ideology of that center.

The seminal significance of this direction and its widespread subsequent representation in other sites in Mesoamerica had long been recognized (Carlson 1981: 159–160). The influence of Teotihuacán had been noted in the ceramics of Central Veracruzan sites. In the

archaeological surveys carried out within the context of the Mata-
capan Project in the Tuxtlas, just to the south and east of our region,
virtually all of the larger communities that could be classified on
various criteria as having arisen under Teotihuacán's influence also
had the Teotihuacán alignment (personal communication, Santley
1983).To extend this influence into an interpretation of agricultural remains seemed no great leap of logic.

The predominant direction also suggested a chronology for the remains. Teotihuacán's epoch is roughly the first seven centuries A.D.
The ceramic evidence to be cited a few pages on fits into the latter
part of that period.

Further caution is necessary here. Teotihuacán's orientation is
recognizable in centers built within the vicinity of its ruins many
hundreds of years later (Carlson 1977: 80). Its influence may have
been expressed in the wetlands of Veracruz a good deal later too,
perhaps with the original reasons for it long since lost.

The association of agricultural features with the latter part of
Teotihuacán's epoch has been strengthened in the investigation of
stone lines on the piedmont west of the Basin (figure 2.22; Sluyter
and Siemens 1992). The stone lines are accordant with the remains
of canals in neighboring wetlands. A slight anticlockwise orientation is apparent in places as well; it may be a suggestion of an earlier organizing principle (Aveni 1980: 236). Its archetype seems to
be La Venta, the Olmec site located on the coast of western Tabasco,
dating back to the first millennium B.C.

The variant is expressed strongly in the Tulipan complex (figures
2.14, 2.15) and relates to a test excavation performed by Mario
Navarrete in 1984 on the "island" of settlement within the platforms and canals. He encountered here among his modest yield of
ceramic fragments a few that were diagnostic of the Pre-Classic and
one that was clearly Olmecoid. He expressed his caution, as anyone would, about such slim evidence; this could easily have been
carried in from who knows where. However, it was a similar faint
Pre-classic signal to the one we were to pick up in the ceramics out
of the test pits in El Yagual (Siemens et al. 1988).

The objective of the dominant orientation, which offsets clockwise from our own cardinal directions, seems to have been the point
on the horizon at which the sun rises on the day of its zenithal passage over areas around 20 N Latitude, which is currently May 1
(Aveni and Gibbs 1976). This passage signaled the coming of the
rains and the time for planting. By aligning major constructions
toward such an objective, space and time could be brought into
accord, allegiance and devotion demonstrated, tradition upheld.

Tichy has demonstrated very well indeed how a similar orientation gave the ancient cultural landscape of upland Puebla and Tlaxcala a remarkable gist, which has persisted into the present time (Tichy 1979).

Such fashioning of the key lineaments of a cultural landscape seems very much like geomancy—the use of a complex system of physical, cultural, and, indeed, mystical elements to place the living or the dead in harmony with cosmic forces (Carlson 1976: 112). And there is some slight evidence that lodestones were used in Mesoamerica to perpetrate propitious orientations (Carlson 1977: 71).

The presence of an apparently quite unecological organizing principle evokes a frustrating reminder. The subsistence system we are dealing with seems to have been environmentally sensitive and effective, in fact brilliant, in practical terms, but it should not be taken as reflecting overly pragmatic decision-making on the part of the builders. In ancient Mesoamerican thought:

> Various aspects of nature [were seen] as integral parts of a single cosmic system . . . Ultimate reality was to be found in the world of the spirit. The material reality of this world, the reality seen as fundamental by modern man, was for that system an ephemeral, kaleidoscopic projection of a mysterious spiritual reality that man could comprehend only partially and express only through metaphor. . . . (Markman and Markman 1989: xix, 65)

Below the Surface

The Sediments and the Anthropogenic Horizon

It was obviously important to examine the stratigraphy underneath the patterning within the wetlands at representative locations (Siemens et al. 1988; Hebda et al. 1991). Test pits were dug into the remains of platforms and canals in two complexes. Soil samples were taken for standard physical and chemical analyses, as well as for pollen and phytolith counts. These last are microscopic silica casts left by some decayed plant cells, which may be diagnostic of species; fortunately, maize often leaves recognizable phytoliths. There is some controversy on phytolith analysis (see, for example, Doolittle and Frederick 1991), which cannot be gone into very far here, but the corroboration of phytoliths by pollen in this case lends weight to the former. Fragments of organic materials were taken for radiocarbon dating. Earth was removed from the pits by regular layers and examined for ceramics. In addition, many probes were carried out with corers of various diameters.

One hoped for some distinct lines on the walls of the pits: a surface into which the canals were cut and onto which the spoil was heaped to form the platforms, perhaps a succession of horizons in the remains of the platforms, indicating use, abandonment, and resurfacing, or, at least, a boundary between a platform and a canal. Nothing like this was visually apparent or indicated by the physical and chemical soil analyses.

A more general stratigraphy could be deduced from the pits and the probes. To begin with, at the base of the sequence that interests us here, about three metres below the surface, is a massive, dense layer of fairly coarse brown sand. Its contact with overlying sediments may be sharp or gradational. This seems to be an intertidal or subtidal deposit like those along the nearby coastline of today.

Marl, a fine-grained deposit including substantial amounts of calcium carbonate, overlies the sand in the southern part of the Basin. It is found in thin layers of no more than 40 centimeters; it probably accumulated in a shallow lake far from fluvial sources.

Above the marl or the sand lies dark gray silty clay. It is about 1.5 metres thick in the south, thins to .5 metres about two-thirds of the way up the Basin, and then fades out in the levee deposits of the northern extremity. It does not show much layering and is penetrated by burrows of freshwater shrimp and by roots.

The silty clay was found to contain ceramic fragments, stone flakes, shells, and numerous flecks of charcoal. Maize pollen and phytoliths were detected as well (Siemens et al. 1988). Pre-Hispanic canals penetrate into this unit but not below. This is clearly a horizon affected by human activity; it has been taken as the horizon to which the surficial patterning pertains and hence the datum plane of this study.

A strong concentration of cross-shaped phytoliths identifiable as maize was found at 90 to 110 centimeters below the surface of the platform remains but not below the remains of canals. They were accompanied by phytoliths of certain wild plants that indicated the maize was grown in a well-drained context. It was possible now to affirm that agriculture indeed had been practiced in the wetlands and that the remains of maize were associated with a canalized surface and conceivably also the unmodified wetlands surface that antedated it.

A chronology became apparent as well. Not many sherds may be expected in an agricultural context, and those that are found are likely to be in poor condition. Nevertheless, several hundred were

Table 2.1. *Ceramic Counts from Remains of Platform in El Yagual*

Levels	Ceramic Types										
(cm)	1	2	3	4	5	6	7	8	9	Total	Percent
1–20	1	—	—	1	—	—	3	9	5	19	8.7
20–40	1	—	—	—	1	—	3	8	6	19	9.2
40–60	—	—	—	—	—	—	—	—	—	—	—
60–80	6	—	—	—	2	2	2	12	9	33	15.9
80–100	4	—	—	1	3	3	9	25	28	73	35.3
100–120	8	1	1	—	2	—	—	22	30	64	30.9
Total	20	1	1	2	8	5	17	76	78	208	
Percent	9.6	0.5	0.5	1	3.8	2.4	8.2	36.5	37.5		100.0

(Source: Siemens et al. 1988: 107).

retrieved from one excavation into a platform (table 2.1). They were concentrated in roughly the horizon of the phytoliths and enough of them were diagnostic to indicate deposition from the Upper Preclassic (several centuries B.C.) to the Early Classic (some five centuries A.D.).

And here one must now place an important parenthesis. The first report of this chronology was audacious, considering our necessarily rough and ready methodology (Siemens et al. 1988). Andrew Sluyter has subsequently cored Laguna Catarina, not many kilometres removed from our own investigations and just downslope, on the heuristic model proposed at the end of the last chapter. His probes are a good deal deeper than ours, in several respects; his analysis has been done with much facility and great care. He presents an "absolute" chronology in place of our "relative" chronology. It was most gratifying therefore to find that his results confirmed our own and elaborated on them (Sluyter 1995: 235–236). Agriculture was found to have been carried on for a very considerable period of time up to and including ca. 500 A.D. in the surroundings of the lake and then faded out shortly after. There is evidence from the Sierra de Los Tuxtlas, southeast of Veracruz, which seems regionally relevant—and corroborative (Byrne and Horn 1989; Byrne, personal communication, 1992; Goman 1992). Forest clearance is apparent for the Formative (Preclassic) period and again, more intensively, in the Classic. Then there is an abrupt decline in agricultural indicators in the Late Classic (1,300 B.P.). In addition to the various parallel indications of Classic–Late Classic wetland agriculture in the Maya area already reviewed in chapter 1, there is now further interesting data from Northern Belize: "agricultural water management by 1500 B.P. at the latest, and possibly as early as 2500 B.P." (Jacob and Hallmark 1996). The investigation of Prehispanic wetland agriculture in the San Juan Basin in Veracruz gave a trans-lowland scope to study of agricultural intensification around and within wetlands.

Going back to our own analysis, two points need to be emphasized regarding the key stratum, the dark silty clay. These qualifications must be kept in mind when one views the patterning. The clays with the organic materials and artifacts had silt added continuously through seasonal flooding. This was worked in by cultivation and canal excavation, as well as other bioperturbation, and darkened by the charcoal left after burning. Moreover the distribution of the phytoliths within a broad and indefinitely bounded band within the column, approximately coterminous with the dark silty clays, does not allow us to deduce that the maize was grown only on the planting platforms. It may also have been grown before that on the seasonally exposed and unmodified wetland margins. So, although the planting platforms and canals fairly clearly represent the intensification of wetland agriculture, they must still be thought of as having been managed subject to the yearly rise and fall of water levels, and to have been preceded, quite likely, by fugitive or flood-recessional agriculture.

Brown silt and silty fine sand occur near the surface along the axial drainage channel and the northern extremity of the Basin. They show well-developed sedimentary structures and contain little or no charcoal. In contrast to the silty clays, the silty sands were laid down in areas of active deposition near stream channels such as those of the canalized San Juan or the La Antigua. The former would be expected to be thicker toward the south, the latter thicker toward the north.

The thirty or forty centimeters below the surface are full of the remains of forest clearing; just under them lies what seems an old soil, but this extends over the remains of platforms as well as canals. Radiocarbon dating confirmed that all this was historic.

Fibrous brown peat with wood fragments caps the sequence in the southern half of the Basin where the test pits were dug. It has been formed relatively recently in the wetlands from the remains of aquatic plants, sedges, grasses, shrubs, and trees. The wetlands have been inundated for sufficiently long periods each year to thwart the complete breakdown of organic matter. This peat layer at the surface has accumulated after the abandonment of agriculture within the Basin.

Associated Settlement

The chronology apparent in the anthropogenic horizon was not surprising. The surroundings of the Basin and the hill lands to the west are rich with the remains of settlement; a number of sites immediately around the Basin and on firm ground within it have been excavated in the past few decades or are known through surface reconnaissance (Navarrete 1981). Domestic and ceremonial architecture was encountered. Small groupings of Preclassic habitations are envisaged along the wetland margins. Larger and more elaborate settlements are apparent for the Classic period, diminished occupance for the Postclassic. Chronological indications in settlement thus coincide roughly with those in wetland agriculture. An approximation of its geography in the Basin and immediate surroundings may be excerpted from Sluyter's painstaking compilation of published references to sites in Central Veracruz and the location of those sites (Sluyter 1995: 143–148, 456–475). However, some aspects of that geography may be more a reflection of the concentrations of recent archaeological efforts than of the actual distribution, such as the grouping of sites along the La Antigua River. Much remains to be done in the elaboration of relationships between settlement and agriculture within this region and, indeed, in the clarification of Prehispanic settlement within Central Veracruz as a whole.

The truly monumental sites are far away from this concentration of "intensive" agriculture. El Tajín lies well to the north and the Olmec centers are as far away to the south. Cempoala is only about ten kilometres to the north of the La Antigua River, but it is also later in its chronology than the patterning. This underlines the growing realization that there is as yet little demonstrable relationship between "intensive" wetland agriculture and statecraft in the lowlands. Lesser linkages are apparent; there were sizable population concentrations near the remains, between wetlands and other types of sustaining areas on neighboring terra firma.

It is probable that the patterning and related settlement in Central Veracruz cannot be attributed to the Totonaca, the people that first come to mind when one considers the prehistory of this region. Nor is the patterning in Northern Veracruz likely to prove Huastecan. Both cultures are later than the chronology apparent from the excavations in the wetlands. Did the builders pertain to the Olmec culture? There is the orientational straw in the wind. The Olmec heartland developed in Southern Veracruz and western Tabasco during "Late Early and Early Middle Preclassic periods— ca. 1200–600 B.C." (Henderson 1979: 83). This is earlier than our chronology.

There were invaders from the uplands into Central Veracruz. García Payon, basing himself on a Spanish chronicler, has them entering from the west from the third century A.D. onward, bringing Teotihuacán influences with them, but reaching only into the Central Veracruzan highlands (García Payon in Ochoa 1989: 235– 238). On his map a blank space extends over the coastal lowland within which wetland patterning is found. We are left with the unresolved question as to just who the builders of this system were.

Analogues

Finally, some contemporary agricultural and, indeed, other activities can also help in the interpretation of Prehispanic vestiges. I have already made reference to chinampas and fugitive agriculture in the development of a heuristic model in chapter 1. The planting platforms and canals, the "raised fields," that are deduced from the patterning are not in the traditions of well-rooted inhabitants, nor of the recent in-migrants, anywhere in the Gulf Lowlands; extensive field inquiry has made that plain. Any help for the interpretation of the "raised fields" will be quite indirect, but not inconsequential.

Historical and contemporary incursions are likely to be instructive with respect to both the impetus and the means by which the wetland is made tractable. The social and economic context of today seems worlds removed from what will have prevailed when the wetlands were patterned, but present relationships between power and access are suggestive, as are those between various degrees of economic urgency and the scale of an incursion. As to the means, the specific techniques and crops are of less interest than the gist of the activity. One needs to sense the seasonal beat and to search out the complementarities.

The strongest current expression of the basic seasonal beat of this landscape is still to be found in the rhythms of ranching. Feral animals responded to it on their own in this Basin before anything that could be called ranching got underway. Subsequently it has been dealt with by transhumance. Much has been expended in recent years to mitigate or even out seasonal differences in the behavior of the water, but all of that does not yet hide the basic beat.

The traditional activities of rooted people have usually been taken as most likely to be instructive of ancient agricultural practice, but the responses of newcomers can be just as interesting. They will learn what they can, or more likely what they are prepared to accept, from longer-settled neighbors, but may also have to respond directly with whatever audacity they can muster and according to their own best lights. Such satisfaction of need within floodplain constraints can be suggestive of initial wetland entries long ago. Something like that may be seen along the gentle slope into a wetland near El Palmar already discussed in the previous chapter (figure 1.11).

Further curious examples of this became apparent during aerial reconnaissance over Central Veracruz in the early 1980s. In the Tulipan and Nevería wetlands there are the striae of machine cultivation (figures 2.14, 2.18). Field inquiry indicated that during the time of the ill-fated Sistema Alimentario Mexicano of the late 1970s and early 1980s, funds were made available to cattlemen for agricultural production on a portion of their pastures. For just a short period in many places within the lowlands, cattlemen carried out a mechanized version of the venturesome catch-as-catch-can cropping of the dry season, or *tonalmil,* on wetland margins. It was a response to a stimulus particular to the economy and politics of the time but suggestive of other possible sudden incursions, such as a need to provide tribute goods in times past. And it also reminded one of what was quite possibly the antecedent to intensification with platforms and canals.

The agriculturalist between microenvironments then and now sometimes resembles a conservative roulette player. He spreads his chips over the layout, some on the squares, where the odds are highest, but some also on the lines between the squares and on down to the spaces where the odds are even and constitute a kind of insurance. He is very different from the player who reaches in with a column of chips and wagers them all on one square, as a modern commercial farmer might want, or need, to do.

To improve one's chances on land subject to inundation—that is, to allow earlier and more reliable access to flooded land, as well as to have water available for irrigation in the dry season—it is useful to canalize. Any recent efforts to that effect, particularly by individual agriculturalists and ranchers, are potentially useful in the interpretation of ancient canalization. Nobody is cutting multipurpose quadrate webs of canals through the wetlands as was done in Prehispanic times. Instead, drainage is served by acutely V'd systems conforming to current ideas regarding drainage. Wells or trenches are dug down to just below the dry-season nadir of the water table in order to provide intakes for pumps that bring water to the land within and beyond the wetland margin. The effects of the latter and the earlier systems are probably not too different: rapid removal of floodwaters after the end of the rains and then subsequent irrigation of the surfaces that have dried out.

A Year in the Basin ca. A.D. 500

From these various lines of direct and indirect evidence one may reconstruct aspect and activity in the Basin during a normal yearly cycle, a period without hurricanes or unseasonable weather, sometime late in the fifth century A.D. The question raised at the outset of the chapter, "In such a place there could be no hunger?" is derived from a legendary account of the yearly round in a place very much like the Basin (Durán, as cited in Melgarejo 1980: 120). The gist of the account is that indeed hunger *was* most unlikely. This cannot be taken literally, of course, nor will it have pertained indefinitely, but the indications of interdigitated production that arise in numerous places in the analysis of the Basin do give one respect for Durán and are the main stimuli for this reconstruction.

Settlement fringes and spots the bottomlands in that distant time; many communities share an interest in the management of the areas subject to inundation. People can be at least seasonally in contact across the Basin. It is a distinguishable socioeconomic unit very probably with the appropriate toponymy.

Looking eastward or westward from one of the "islands" of firm ground shortly after one or another of the two precipitation peaks in the wet season, one sees green hills in the distance. Crops are well along on the fields that have been cleared for this year's wet-season cultivation; patches of low forest and grassland remain— the annually revitalized savanna.

The foreground is a watery world. Planting platforms on the wetland margins are mostly submerged, undergoing fallow. Much of the land subject to inundation has been cleared, leaving clumps of tropical forest or its succession species; hydrophytes thrive in the shallow areas. This vegetation protrudes above the surface of the water to varying degrees; gallery forests on levees divide the lake with colonnades. The inhabitants on the margins or on the "islands" can travel freely by canoe, but must wend their way.

This is the time of much trade and socializing. The wet-season crops of the hill lands are becoming available. It is also the time of wide views from the water's surface, particularly on the margins of the full, extended Laguna Catarina looking northward. The islands with their structures are mirrored dramatically.

There are swarms of bothersome insects, and one does wonder how the inhabitants coped—no doubt taking advantage of smoke and breezes. However, the vectors of malaria and yellow fever have not yet been brought in to plague them.

With the onset of the dry season the upper, peripheral planting platforms emerge and are planted, then those downslope; one can envisage staggered cropping and commensurate maintenance of the field margins. There has been slumping, as always, around the edges during the past wet season.

An observer looking about in March or April, near the height of the dry season, sees maize and other crops varying in maturity from the slightly higher surfaces of the margins to the lower ones near the centers of the wetlands. There are stands of palms within these fields, much as one sees them today within the pastures. Canoes move through the labyrinth of waterways, carrying workers, harvested crops, or perhaps loads of muck scooped from the bottoms of the canals and intended as fertilizer.

In some places any as yet uncanalized exposed surface may be cropped as is, simply to try one additional thing before the next *cresciente,* the next rise in water levels.

The height of the dry season is also the time to secure areas like this by means of canalization, by the expansion of the system of planting platforms. A surveyor-priest leads the way, sighting, staking, and performing the necessary ritual. Gangs of workers continue on as long as the rains hold off, with their adzes and baskets, laboriously treading the sucking mud, enlarging the web.

One can hunt terrestrial herbivores at various times of year in various areas, but the dry season is a good time to hunt out on the wetlands, perhaps for the ducks that have migrated southward at the onset of the winter in northern areas, perhaps for other aquatic fowl as well. It is also a good time to fish. Prevented by natural or artificial means from returning to the main stream, fish are now concentrated into pools or in the canals between the fields and have grown fat. Groups of men join in fish drives along canals toward transverse nets, making of it all a social occasion.

Most people live along the margin of the basin on slightly higher ground. Shade, fruit, and ornamental trees provide a green matrix for houses of pole and thatch and for the sacred buildings on their mounds. Household gardens help to sustain, of course, and furnish items for trade. Some household animals are about as well; they are not as many or as varied as they will be after Contact.

On the hills to the west, behind the settlements, an observer sees patches of dormant gray brown forest and harvested fields, still showing the refuse of the last wet-season crop. Fires sweep the horizon, removing the hardened remains of grasses on the savannas. After new shoots have sprouted it will be a good time to go after deer. Nearer by, people are preparing land for a new wet-season crop.

In places out on the sloping hill lands, stones have been gathered into rows, some in one direction and others at rough right angles. A surveyor-priest has been active here as well, preserving the sacred direction, but trade-offs have had to be made. The lines must be religiously correct and at the same time mostly transverse to slope so that erosion is retarded during the rains and soil as well as moisture are preserved behind the rows. Maguey may be planted on the rows to aid the retention and provide an additional resource. Some fields are taken up by tree cotton, in others two crops of maize may be obtained during the year. By these means rainfed agriculture has been intensified.

Some of the dunes to the east are in motion during the dry season, especially when *nortes* strike. On others there is vegetation, equally gray still, as on the hills to the west, but lower, more tangled and thornier. This has always been difficult to penetrate. Some of this land has been used for wet-season agriculture and is filled now with weeds and crop debris. Looking downslope from the east or the west, the bottomlands seem an oasis.

On the gently sloping alluvium within the Basin, around and to the north of Caño Prieto, for example, it is possible to grow a good dry-season crop on land not subject to inundation but with the residual moisture from seepage still in the soil. The hollows adjacent to the mounds begin the dry season full of water—an important resource, at close proximity to settlement; this goes down in the succeeding weeks, leaving small, moist surfaces for some further carefully timed cropping.

By this time the priests are observing the heavens, carrying out their rituals and making their pronouncements about the coming of the rains. When they do come the hills green, and agriculture there resumes. In the wetlands the fields are soon submerged. Harvesting on the lowest and last exposed fields may have to be done from canoes; one just may lose that last crop. The Basin has become a watery landscape once again.

3. Probing the Ethnohistorical Literature Surrounding the Encounter

Todo la costa despoblada.

—after Gerhard 1986: 375

From excavations to documents, from prehistory into history—many of those investigating Prehispanic wetland agriculture have felt it necessary to make this move. We have searched the accounts of the Spanish *entradas* in order to probe the "protohistoric" period (Adams 1977: 258) for anything that might bridge the gap between the time of the use of the field systems and Contact, as well as to appreciate fully those first views and to gauge the early impact. The scholarship has been diligent but the results meager.

Not too long ago the Encounter was reconsidered at length and from countless perspectives, including the geographical (e.g., Butzer 1992; Turner et al. 1995). We will not have to go through all that again for at least another ninety years! Only certain aspects interest us here and they can be briefly put.

Hiatus in the Lowlands of Central Veracruz

The ceramic and microfossil evidence that suggests a chronology of wetland agriculture fades upward quickly from the horizon associated with the patterning. Dark silty clay continues, grading into peat interspersed with roots and other forest debris (Siemens et al. 1988). Intensive use of patterned wetlands lapsed in the Classic period, as much as a thousand years before Conquest.

A good deal is known by the archaeologists of the region—particularly those who have had their base in the Museum of Anthropology in Jalapa—about Prehispanic settlement in Central Veracruz, but not much of what is known has been published. Also, the phases of the cultural chronology of the Veracruzan lowlands are still not very clear, and the Post-Classic period is generally not well understood. However, there are indications that the settlements immediately around the wetlands were abandoned at roughly the same time as the intensive wetland and piedmont agriculture (Navarrete 1983).

When maintenance lapsed, for whatever reason, canals and planting platforms will have deteriorated quickly. The water rose and fell, surfaces rounded, siltation continued, tree growth and hydrophytes thickened. Vegetation itself will have distanced the wetlands from whatever settlement and agriculture persisted on the surrounding hill land. For many years few but hunters and fishermen will have ventured into the wetlands; although not far away, they will have been the back of beyond.

It is quite possible that a less intensive seasonal use of wetland margins, without canalization but with links to adjacent microenvironments, was being carried out in many places during the centuries before Conquest. And it could be that in some wetlands within the Gulf Lowlands canalization and field build-up was still going on at Conquest. But it does not seem likely in Central Veracruz.

Sluyter was able to find an intriguing late-sixteenth-century reference in this connection. Someone was inspecting land requested for a livestock *estancia* and came on a patterned wetland just south of the port of Veracruz: "A lagoon forms during the rains . . . and marshes with straight channels running in a southerly direction" (Mercedes, cited in Sluyter 1995: 201; my translation). It seems fairly clear that these are long-abandoned vestiges not too different from what a careful observer might see in a Central Veracruzan wetland today.

Encountering Productivity

Bernal Díaz del Castillo's recollections of the landfall remain absorbing reading (1963). The old soldier, who began his account when he was over seventy, felt he needed to set the record straight, and he did indeed become the prime source on the Conquest of New Spain. His memory seems to have been excellent. His self-deprecating asides in excuse for this or that lack of detail are delightful: they were young, he points out, and often preoccupied; and they were soldiers—as if to ask how they could also be expected to observe carefully. The observations of landscape in his account, in that of López de Gómara and in Cortés's letters, too, are in fact meager. The precise wordings cannot be pursued very far: the observers were applying medieval categories and were indeed preoccupied; however, when various indications are brought together, they do put all the lowlands in a certain light.

There is a good deal about the coast, the currents, weather conditions, anchorages, the sandy shoreline, the insects, and the scarcity of food on board. Arriving opposite the Basin, the Spaniards appreciated very much what was brought as gifts and for barter from nearby settlements. On their first foray toward Cempoala, they were impressed by the greenery that surrounded it (Díaz del Castillo 1963: 107). López de Gómara was even more enthusiastic: within the town itself it was, "all gardens and greenery and well-watered orchards" (1964: 70). It was no wonder that they should have been impressed by greenery. They arrived just before Easter, at the height of the dry season, when the vegetation on dunes and hill land is gray and dormant. There is very little more. Even after they had chosen the site of the present La Antigua as the place to build Veracruz, there is nothing explicit about the Basin that began just on the other side of the river, where most of the greenery was.

One searches for comparative and perhaps elaborative materials in the descriptions of first encounters in the lowland arc around the Gulf, and this can begin with Acalan, the Maya province at the western base of the Yucatan Peninsula, already discussed at some length in chapter 1. Cortés came by it on his remarkable overland journey to Honduras in 1524–1525. The visit and the place emerge from a suite of documents found and analyzed by Scholes and Roys (1968). The result is a rather good portrait of a place, just before most of its people were swept away. It is effective, unselfconscious geography from an historian and a linguist.

Two hundred thirty soldiers, 93 horsemen and 3000 unwilling Mexican participants under the command of native chieftains, including Cuauhtemoc, made their way across the grain of the lowlands, slogging through wetlands and building many bridges. This seems as wrongheaded as it was heroic. What a costly and inappropriate way to move through such terrain! By the time the company approached Acalan they were desperate for food, and they found it, in abundance. Subsequent tribute lists indicate something of the variety of what will have become available: honey, hens, beans, maize, squash seeds, chiles, and calabashes (Scholes and Roys 1968: 396). Cortés and his companions were to extol the wealth and the resources of Acalan, but they did not find gold, which is not surprising since the streams of the province emerge from karstic terrain.

Nothing is said about the system by means of which the abundance was produced, and there is certainly no mention of anything like the platforms and canals. One might think that if they had been in use, the Spaniards, who had already been to Central Mexico where a similar system was in operation, would have noticed.

However, in the account of their arrival in Tenochtitlán there is a good deal about their enchantment over gardens, canals, and canoes, but not an explicit word there either about the method of cultivation, even though they were in the midst of chinampas in full production (Díaz del Castillo 1963: 214, 231). They were soldiers, and preoccupied, after all.

In several sources to canalization in wetlands along the Usumacinta River, there are intriguing references that have been interpreted as indicating the use of planting platforms and canals at the time of Conquest (Pohl, ed., 1985: 36). Moving westward in the lowlands of Tabasco, occupied by the Chontal Maya, one comes on the brief note by Cortés: "The land is very fertile and abounds in maize, fruit and fish and other things which they eat" (Cortés 1972: 23).

The conditions in Tabasco at Contact, and a great deal more, are outlined very well in a comprehensive and often incisive regional geographical study by West, Psuty, and Thom (1969: 89–103). They review early Post-Contact Spanish accounts, which portray the Chontalpa as quite densely populated, well connected eastward and westward by overland and coastal trading routes, and as agriculturally very productive. Maize and a range of other food crops are noted—grown presumably by means of slash-and-burn cultivation—as well as an export crop, cacao. Moreover:

> A common ecological pattern underlay the subsistence base of practically all the Tabasco lowlands: fertile soils of the stream levees, abundant and varied aquatic life in the rivers and adjacent marshes and lakes, and plentiful game within the savanna grasslands and *acahual* (abandoned slash-burn fields) nearby. (West, Psuty, and Thom 1969: 101)

This may be taken as a good schematic, micro- environmentally articulated summary of subsistence in the lowlands generally. In fact, the complex, seasonally varied patchwork of mid-twentieth-century lowland Tabasco, before it was affected by the exploitation of petroleum and modernization, is a good approximation of what would have been seen in the floodplains of the lowland rivers generally at Contact. These floodplains have been searched for the remains of intensive wetland agriculture, but the characteristic patterning has not been found in any substantial complexes. The system may well have been unfeasible and unnecessary, as was pointed out in chapter 1.

Agricultural productivity emerges from depictions of what was the Olmec heartland: Western Tabasco and Southern Veracruz, the base of "the first truly complex culture in Mesoamerica" (Henderson 1979: 89). The culture pertains to the Preclassic and antedates most of the evidence of wetland agriculture in the Mesoamerican lowlands, and indeed no traces of patterning were found in the wetlands during our own air reconnaissance over the area. By the time of the Conquest, the Olmec heartland was occupied by Nahuatl and Popoluca peoples (West, Psuty, and Thom 1969: 99). These lowlands yielded rubber, cotton, cacao, and every sort of food, as well as various articles of luxury. Aztec merchants' trade routes passed through the region (Scholes and Warren 1968: 776).

One of Cortés's captains, Diego de Ordáz, was sent from Tenochtitlán to the lands around the lower Coatzacoalcos in order to assert Cortés's authority there. He found the mouth of the river navigable; here was a good port, but it was too distant from the capital to be useful. However, there was good farming and grazing land (Díaz del Castillo 1963: 268).

Just how productive and sustaining these particular lowlands have been for millennia emerges very well from Coe and Diehl's remarkable book (1980) on the people of the lower Coatzacoalcos in the 1960s. They sketch interdigitated agriculture on the various microenvironments of the floodplain, stressing the importance of the cultivation, without canalization, of the seasonally emergent, gentle backslopes of the levees, which complements cultivation on neighboring hill land. This, they believe, plus all the other productive activities carried out in this environment, was more than sufficient to sustain the population numbers that may be deduced for the Olmec period. Such a system was probably what the Spaniards saw in action too. It is sobering to see how this setting has been polluted by the petrochemical industry in our day.

Continuing westward: Pedro de Alvarado entered the Papaloapan River Basin without the permission of his commander, Juan de Grijalva, but is memorialized to this day in the name of the town at its mouth. On this and the subsequent entry during Cortés's journey westward along the coast, some of the settlements of the Basin were explored and productivity found here, too. López de Gómara declared, "The land it drains is good land; its banks are beautiful" (1964: 52). He went on to catalogue an amazing array of fish, birds, and land animals, but he also noted the dramatic flooding for which this Basin would become notorious. It would be the venue of one of the great Mexican river-development projects of the mid-twentieth century, the prime imperative of which was flood control.

Stark has looked very carefully at the meager evidence on the economy of the Papaloapan Basin at Contact and has identified

some specialization among the communities in the various physiographic regions and a location of the larger ones between contrast-
ing resource zones (Stark 1974, 1978). Something like this is plausible for other river basins of the lowlands at the time; it is certainly
indicated for the surroundings of the San Juan Basin. We have come
to appreciate such locations as suggestions of effective exploitation of complementary biomes.

Looking northward from Central Veracruz and reviewing early
reports on the basins of that part of the coastal plain yields us nothing very useful out of the accounts of the conquerors but does bring
us to the impressive work of subsequent analysts Kelly and Palerm
(1952), and also that of Stresser-Péan, particularly the latter's review
of the ancient sources on the Huasteca, as this region has long been
known, and most particularly to one statement:

> The Aztec applied to the Huasteca, as to other warm moist
> lands of the Gulf of Mexico, the name Tonacatlalpan, "land of
> food," doubtless being of the opinion that these regions were
> fertile and free from drought. They also thought that the name
> was justified by the great variety of plants cultivated in the
> warm parts of the Atlantic slope. (Stresser-Péan 1968: 588)

So, although one must postulate Preconquest agricultural abandonment in at least some of the Gulf Lowlands once used intensively, the general aura of these lowlands at Conquest is one of
productivity. It is not at all clear how these conditions are to be
reconciled. We have only a very inadequate understanding of the
phasing and regional variation in occupance and land use in the Post
Classic. There were sizable populations in the lowlands, generally,
at the time of Contact, but how they were sustained in any given
area is not so apparent. A variety of linked rudimentary production
in neighboring microenvironments, without agricultural "intensification," could quite well have been sufficient to give the lowlands
their aura. Perhaps, already at this early stage, the documents are
showing us a kind of basal, and at least potentially bountiful, subsistence production that will be a strong undertone to discussions
of commercial production in the documents of later periods.

It may be a matter of representation. Greenblatt's *Marvelous
Possessions: the Wonder of the New World* (1991) comes to mind.
The productivity seen in tropical lands seems to have been an important part of the wondrousness of the New World in the early accounts. To show this world thus reflected the strain that its discovery put on inherited concepts and may have been a recognition

of difference. It also corresponded to the needs of the conquerors and their sovereigns; it served the imperatives of colonial appropriation. Stressing tropical lowland productivity serves this analysis as well; the reader must beware.

Taking Possession

Díaz del Castillo was critical of the substance and the style of López de Gómara's account of the Conquest, and one tends to accept this judgment, but it is difficult to resist dipping into the denigrated account for some of the words put into Cortés's mouth. The conquerors were together at their first settlement in Central Veracruz. They had by this time explored the lowland coast and made a number of incursions; Cortés considered it was time for some formalities:

> He called them together and spoke to them, saying that God had evidently favored them by guiding and bringing them safely to such a good and rich land (as it truly seemed to be, to judge by what they had seen in the short time since their arrival), abundant in provisions and filled with people, who were better dressed, more civilized, reasonable and intelligent, with better houses and farms, than all the others thus far seen or discovered in the Indies; that it was likely there was much more to it than they had seen; that for that reason they should thank God and make a settlement there, and penetrate the country and enjoy the grace and favor that Our Lord had shown them. . . . So saying, Cortés took possession of the country. (López de Gómara 1964: 65–66)

Impact

The demographic impact of Conquest is common knowledge in general terms. When one reviews the regional estimates of depopulation presented by Gerhard (1972), by West and his colleagues (1969), and by Scholes and Roys (1968), and then assembles a map such as that in figure 3.1, the impact has an impact again. Between 1520 and 1570 over 90 percent of the population of the lowlands died.

The massive literature reviewed in these and other sources indicates various causes. In some areas the lowland population was reduced by slaving and warfare, as in the Panuco region. In other areas there were particularly severe abuses by *encomenderos* and

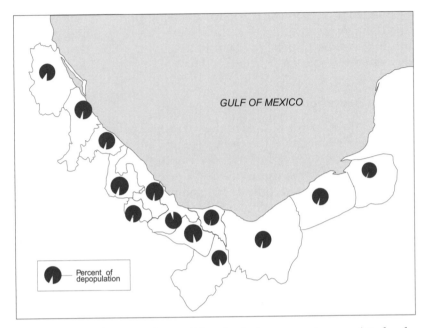

Fig. 3.1. Depopulation in the Gulf Lowlands from 1520 to 1570. (Gerhard 1972; Scholes and Roys 1968; West, Psuty, and Thom 1969).

disruption of trade, as in Acalan. Cattle menaced Indian agriculture and may have been a factor in depopulation within the Chontalpa (West et al. 1969: 118). However, the primary factor throughout the lowlands was decimation by disease. Alba González Jácome has traced the specific diseases that were involved and reviewed the grim dimensions of this whole catastrophe (1988: 4–23).

It is easy to imagine the vegetational changes that might ensue in the lowlands, but in order to get beyond generalities there must be a recognition of at least the principal physiographic variations within them, moving from the sea inland: beach ridges and dunes; floodplains with levees and wetlands underlain by recent deposits; hill land underlain by older sedimentary formations, eroded and trenched—all this interrupted by the volcanic hill and mountain massif in the Tuxtla region. The picture is complicated further by climatic variations: an increase of precipitation from the coast southward toward the mountains, and semiarid enclaves in shadowed areas, particularly in Central Veracruz.

The areas that were found to be well populated and productive at Contact can logically be thought of as having been deforested, or "open" (Parsons 1989: 316), but only in the way that a tropical terrain largely subject to shifting cultivation with some sort of fallow can be considered open. Woody growth in one stage or another of regeneration is always in sight. Hill land and the fixed dunes will have had the moth-eaten look that tropical lowland used in this way has now. It is likely that the lowlands proper were mostly covered with forests and that hydrophytes prevailed in the lowest areas.

It is clear that substantial and, indeed, rapid reforestation followed depopulation in the lowlands generally and elsewhere on the mainland, as on the Caribbean islands (Parsons 1989: 316–317). However, this too must not be overgeneralized. The survivors will not have been able to maintain cultivation on more than a portion of even the best lands: the levee tops and other alluvium just above seasonal flooding. The margins of the seasonally flooded wetlands, land from which one can get an additional crop if one needs it and is so inclined, will certainly have been left, if, indeed, any were still in use at all. The newly abandoned land will have been overtaken soon by woody growth—this is hot and humid country. On hill land and on the fixed dunes patchy *acahual* from previous cultivation will have been free to mature back into a low, scrubby tropical forest that loses a good deal of its foliage in the dry season. Kitchen gardens will have been quickly overgrown. The wetlands proper, patterned with the remains of ancient cultivation, will probably already have been covered by forest and hydrophytes for centuries. The forest in these areas must not be thought of as *selva alta* entirely, but rather as a mixed and discontinuous combination of brush, stands of palms, and high tropical forest proper. And there remained areas of open grassland on the hill land, edaphically controlled or anthropogenically produced, often referred to in the early sources as savannas. The Cortés party saw deer on meadows west of what is now La Antigua (Díaz del Castillo 1963: 106).

The abandoned land was left free, as is often said, for the entry and spread of cattle. Much land very suitable for cultivation would thereafter be inaccessible to agriculturalists. The forest debris and the buried soil up near the top of the profiles in the test excavations within the San Juan Basin indicate colonial clearing in aid of ranching within the wetlands and possibly on neighboring terra firma. It may well be that there was soon overgrazing within the newly established cattle holdings, but it is apparent—Sluyter has put it well (1995: 294, 306)—that in this region at least there would not be the landscape degradation during the succeeding colonial centuries that is associated with the Classic period of the Prehispanic era.

4. Reformatting Sixteenth-Century Documents

"Muchos toros y muy bravos."
—Arias Hernández 1571: 199

In March of 1580 Lic. Alonso Hernández Diosdado, an impressive young man, is complying with an order for a *Relación Geográfica* from old Veracruz, the present La Antigua. Before he has gone very far he sketches the San Juan Basin and gives our analysis new impetus:

> Just in front of Veracruz the river is joined by a substantial and surprisingly deep stream. Seven sizable rivers, from sources 10 or 12 leagues away, have already been united in this stream not far above the confluence. In many curves they water the lands and pastures of this district—a most agreeable, beautiful scene. Most of the land is a plain and from any high point one can see with great pleasure how the various channels cross it, watering the land most abundantly and giving it freshness and a lasting green, so much so that it enjoys a perpetual spring. In consequence the district of Veracruz is so fertile and abundant in pasture that within a radius of seven leagues are normally grazed more than 150,000 head of cattle and mules. In addition, innumerable sheep and goats are brought down from Tlaxcala and Cholula and other areas each year to winter in this district. Neighboring provinces resort to it for pastures; in this respect it is the *Extramadura* of the realm. (Hernández Diosdado 1580: 313–314; my translation)

This well-known statement, often cited when early ranching in the New World is discussed, certainly takes a gratifying sweep over the Basin landscape and represents it as favored indeed, but also sounds more than a little ironic, climate-wise, to anyone who has spent time in tropical lowlands. It does indicate that a succession has taken place; the Basin is occupied and is being used again.

Cattle are not the whole succession story, of course; other animals and plants have been introduced. New ways of working and getting rich have been initiated and a new web of settlements and transportation routes thrown over the land. Moreover, new sources and a new framework have been introduced into this analysis. But, cattle epitomize it all, especially the *cimarrón*, the domesticated animal that returns to the wild and appropriates the wetlands.

The Documents

Documentation seems wonderful when one is attempting to reconstruct a prehistoric landscape without it. And yet, once one emerges into history, the task of tracing occupance and use seems not to have been facilitated so much as complicated. It *is* absorbing.

Declaración *and* Relación

Two documents are basic to this reconstruction. The first is one of the reports solicited from parishes in the New World in 1571 (Butzer 1992: 556). Paso y Troncoso, who made it accessible, presents it both as *Apuntes para la descripción de Veracruz* and as *Declaración sobre Veracruz* (1947: 189–201). The declaration was made in Madrid by the priest and vicar of Veracruz, Arias Hernández (1571).

The second is the *Relación de la Ciudad de la Veracruz y su Comarca.* It comes from the first series of *Relaciones Geográficas* (1579–1586) and is usually associated with the name of the *alcalde mayor* of Veracruz at the time, Señor Alvaro Patiño de Avila, from whom it was required. It was actually compiled by a *vecino* of Veracruz, and *médico*, Lic. Alonso Hernández Diosdado (1580).

There does not seem to be any biographical information on Arias Hernández, the vicar, so one tries to find him in his words. It is apparent that someone was recording his statements but that he had the opportunity to supplement them with notes in the margin. He seems laconic, but that may have been the style of his recorder. He has the most engaging way of stringing together what seem at first to be non sequiturs but on reflection become related. He does seem just a bit weary after his time in New Spain. He had obviously observed that now-distant tropical town and its surroundings carefully and no doubt had listened at length to many of its inhabitants. Life was difficult there, but one could eat well.

Alonso Hernández Diosdado is recognized in an introductory letter by the *alcalde mayor* as someone who was well versed in geographical matters (Pasquel 1958: 182). He also, clearly, had observed well; he was ordered in his replies but not always in accord with the categories imposed by the fifty questions to which all those preparing *Relaciones* were to respond in the order prescribed (Cline 1972: 233–237). That alone is a tribute to his thoughtfulness. His prose is fluent and sometimes quite enthusiastic; he seems particularly accomplished in matters nautical, in his characterization of the coast and the winds.

René Acuña, editor of the *Relaciones Geográficas* of New Spain, wonders about the médico (1985: 302–303). The material of the *Relación* does seem to have come from him; however, he may not have been a *vecino* and not yet a médico, nor possibly even a *licenciado*. And he left Veracruz soon after having written the *Relación*, looking for a more agreeable place in which to live. But, one is led to protest, he does seem to have read some medical classics, he certainly uses language carefully and he has produced a rich and careful compendium, so he deserves a certain amount of respect.

The médico seems to have had the *Declaración* of the vicar at hand. The one seems to be drawing on the other, indirectly. The vicar is more detailed in places than the médico, but the latter often has a wider view. Perhaps the two just observed well and had the same informants. In any case, the two documents are excellently complementary. The two authors have different interests and abilities, and are at two different stages in their careers—the vicar being debriefed on his return from a considerable time in the field, the médico still a young man but already in possession of special knowledge; they serve the court, and us, quite well.

Both the *Declaración* and the *Relación* reply to sets of questions, the first implicitly and the second explicitly (Cline 1972: 189, 233–237). This is geography in the service of government; it was to facilitate missionization and colonization. It was generated largely through the initiative of the remarkable Juan López de Velasco, for a time cosmographer and chronicler to the Council of the Indies (Butzer 1992: 554–557). He developed the framework for a regional geography of the New World, first soliciting reports from parishes on a list of items, such as the *Declaración* by the vicar, then formally setting out the questionnaire on fifty items that was dispatched to magistrates and in response to which some five hundred *Relaciones Geográficas* were composed.

Although the questions may be a sensible and perhaps even an inspired list of things that a king should know about his realm, they are still a list, conceived at a desk, something of an affront to those on the scene and faintly idiotic as forms always are. Our respondents find some of the questions unanswerable because they go beyond their competence, or just baffling, raising issues that they have obviously never thought about. Sometimes the questions are answered with cross-references because they seem repetitive or in the wrong order.

The order is curious for us as well and can hardly be used as a template. More importantly, the questions themselves are uncomfortably categorical and dichotomized. It was observed in the 1970s that the *Relaciones* were still little known and rarely used (Cline 1972: 183; West 1972: 396). This is not difficult to understand. The documents are easily excerpted, as in the collection of materials basic to the map of population decline in coastal Mesoamerica presented earlier (figure 3.1), or in the elaboration of any number of possible studies of colonial resources, institutions, occupations or what have you. They are also easily anthologized; segments of both of our main documents are included, more or less intact, in various histories of Veracruz (e.g., Trens 1947: 197–210; Ramírez Cabañez 1943: 17–41; Melgarejo 1981: 121–122). To fully appropriate these documents is another matter; to use them in the reconstruction of a landscape certainly requires much more intrusive handling.

The data are profuse and quite delectable but require reassembly according to our own integrating concepts, such as that of the core *vis-à-vis* the periphery, ecosystems, and altitudinality. There is the danger, of course, that the integrity of the fragments will thus be compromised. It is hoped that this can be minimized by careful regard for context, as though one were undertaking an archaeological analysis. And we are, after all, in a position to read into the many bits far more than the authors dreamed of. More than justice is quite possible.

Repeated scrutiny and endless shuffling against the background of field experience eventually brings considerable empathy. One finds oneself at the same viewpoints, feels the suffocating stillness at noon in August, hears the bustle of the riverside and the street after the arrival of the ships. Such immersion induces the present tense.

Complementary Travelers' Accounts

Various English travelers come to New Spain early in the colonial period. Hakluyt gives us access to the observations of Robert Tomson, merchant, arriving in 1555; M. John Chilton, "desirous to see the world," landing in Veracruz in 1568; and Henry Hawkes, merchant, setting down his observations about travel in New Spain in 1572 (Hakluyt 1927: 246–280).

Two clerical reports are relevant as well. Fray Alonso Ponce, a

Jesuit *visitador* to New Spain in 1584, makes some very usable observations on both San Juan de Ulúa and Veracruz (Pasquel 1979: 225–236). In 1609 Fray Alonso de la Mota y Escobar, bishop of Tlaxcala, journeys through the bishopric, which included lowland Central Veracruz, and provides us with valuable sequels on a number of points (González Jácome 1987).

Pinturas?

The questionnaire that structures the *Relaciones Geográficas* requests locational information *"en pintura"* (Acuña 1985: 18–23 [Items 10, 42, 47]). The maps that do accompany many of the late-sixteenth-century *Relaciones Geográficas* are thus generally referred to as *pinturas*—paintings. They form a vast and most intriguing corpus (Robertson in Cline 1972, 243–278); indigenous cartography was commonplace in the early colonial Americas. The Instruction sought to co-opt it for the *Relaciones,* and, indeed, many of them are accompanied by admirable examples. Some Indians were able to adopt European cartography and use it as a tool of their resistance (Harley 1992: 524). An analysis of the subject can easily bear one off into questions of aesthetics and representation. We have little opportunity for that here.

There are no maps with the *Declaración.* Two maps usually accompany the *Relación,* but they are not terribly helpful. Their scales are quite small, they do not include many of the places mentioned in the *Relación,* and include others that are not. The hydrography of the district of Veracruz was certainly not yet well understood. San Juan de Ulúa is on the mainland when it should be on an island. The Indian communities are roughly where one would expect them to be, but there are more of them than the médico notes in his *Relación* (When the maps were drawn depopulation had not yet proceeded as far as it had at the time of the writing of the *Relación?*) and the names are different.

These maps clearly present a problem. Acuña argues plausibly that they did not accompany the Veracruz *Relación* when it was submitted, but were prepared between 1545 and 1554 and only added to the *Relación* sometime between 1596 and 1611 (Acuña 1985: 301–308). Our two main documents are thus without their *pinturas.*

In another cartographic respect, however, Acuña is unfortunately quite wrong. He points out that the maps could not pertain to the *Relación* because they show Veracruz along the La Antigua River when it had already been moved to where it is now. The locational references of the *Relación* itself, indeed its whole geographic fabric, are nonsense if one does not place Veracruz on the north bank of the La Antigua River. Moreover, independent analysis corroborates the shift from the one location to the other after the *Relación* was prepared (e.g., Rees 1976: 28–30). It is obviously quite possible to trace the paternity of a *Relación* and to edit the text without actually appropriating the contents.

There are at least seven maps with the early colonial land-litigation documents that are roughly relevant chronologically and areally (AGN 1979). They are particularly eloquent as expressions of the taking of possession.

Nature Out of a Classic Mold

In his treatise "On Airs, Waters and Places," Hippocrates advises a physician coming into a strange place:

> To consider its situation, how it lies as to the winds and the rising of the sun . . . These things one ought to consider most attentively, and concerning the waters which the inhabitants use, whether they be marshy and soft, or hard, and running from elevated and rocky situations, and then if saltish and unfit for cooking; and the ground, whether it be naked and deficient in water, or wooded and well watered, and whether it lies in a hollow, confined situation, or is elevated and cold; and the mode in which the inhabitants live, and what are their pursuits, whether they are fond of drinking and eating to excess, and given to indolence . . . From these things he must proceed to investigate everything else. (Hippocrates 1939: 19–20)

Early in the list of questions that elicited the *Relaciones Geográficas*, respondents are asked to characterize the climate of their surroundings and then to describe other aspects of the physical environment in a similar vein:

> 3. Describe generally the climate and quality of the province or district: whether it is very cold or hot, humid or arid; with much or scanty rainfall and when it is greater or lesser; how violently and from which directions the wind blows, and at what times of year.

4. State whether the land is level or rough, open or wooded; with many or few rivers or springs; abundant or lacking in water; fertile or lacking in pasture; abundant or lacking in fruits, and in means of subsistence. (Cline 1972: 234)

Both the Classic and the sixteenth-century formulations represent a shrewd inquiry into the relationships between the physical environment and well-being. The breadth of vision is remarkable. However, the categories and the dichotomies of both the "Hippocratic Heritage" (Sargent 1982) and the sixteenth-century response present problems to twentieth-century readers. We tend to ask for measurement, think in terms of scales, and have some systematic understanding of process. The observations offered about natural phenomena in the *Declaración* and *Relación* often make it seem that we, the sighted, are watching the blind.

It is useful to keep in mind Glacken's observation about the Classical formulations cited, "there is no systematic body of theory here" (Glacken 1967: 93). That assessment seems appropriate for the fifty questions, also.

Airs

The vicar puts the basic climatic facts straightforwardly. He has the rains begin in May and continue until October. Sometimes it rains for weeks on end, but there can also be substantial intervals without rain. And when it rains it rains hard. With the rains come high temperatures; the sun burns and the soil is heated. From October onward, the *nortes* temper the weather; it is possible then to get cool water to drink (Paso y Troncoso 1905 [1947]: 197–198).

The médico is good on "airs"; he does seem to have had a special interest in sailing and in ships. He takes the reader around the compass. The winds are not fully reliable but there is a pattern: between the fall and spring equinoxes, more or less, winds blow from the north. Everyone who writes about lowland Veracruz, it seems, must sooner or later try to give these southward bursts of polar continental air their due:

They blow with a force and violence that words can hardly describe. They do not blow incessantly during the six months of their season but they do come often and impetuously enough that one lives during that time in constant fear of them. (Hernández Diosdado 1580: 311; my translation)

The médico fusses a great deal with shifts in direction. One recognizes the onset of the blander trade winds blowing from the south or southwest in April, eventually bringing the rains. During the summer offshore breezes, called *terrales* by the sailors, set in well before dawn—an exchange of air from the cooled land toward the sea. Onshore breezes begin before midday, especially when the skies are clear; they counter somewhat the fierceness of the sun. Without them it would be difficult to live in such a place, since "This district is hot and very humid; as a result it is unhealthy and full of putrid and dangerous diseases" (Acuña 1985: 311; my translation).

Here is the fundamental indictment of Veracruz—of a piece with judgments of other tropical ports and, indeed, of tropical lowlands in general. Observers coming to the Veracruz coast would remark on the heat and humidity, and the unhealthiness of places, again and again until late in the nineteenth century, and echoes would persist even after the germ theory of disease was common knowledge. Here is one of the major aspects of the colonial assignment of "otherness"; it can be articulated in transoceanic and latitudinal, as well as altitudinal, terms. These places are sick.

One wonders, further, if the lords and agents of upland Teotihuacán and later Tenochtitlán considered these lowland regions of their hegemony in comparable terms. Were they "sick" then, as well? As has been outlined in chapter 2, it would seem that they were not plagued by yellow fever and malaria; the vectors had not yet been introduced. But other forms of denegration are quite likely to have spiced highland-lowland designations, as they have in historic times.

The médico wrestles with the unhealthiness, with questions of cause and effect (Hernández Diosdado 1580: 191–193). His syntax winds and doubles back like the vapors themselves, his synonyms for putrescence grade into each other. The basic point is that heat engenders exhalations from the swamps and the wet sands which, as the vicar points out in simpler words, become fiery hot during that time, so that no one goes there. The vapors collect over the town and threaten its inhabitants.

How? Here the médico shows something of his education and resorts to ancient wisdom, to the heritage of Hippocrates and the Roman physician Galen, too. The cosmos and living beings, and indeed the "atmospheric constitutions" of particular places, are made up of four elements; they correspond to four qualities, or humors. Harmony or balance among the humors means health; an excess of one or the other, which might well be brought about directly by weather, means disease (Sargent 1982: 12–15, 49–52, 68).

The médico is preoccupied by vicissitudes. He finds that the onset of the rains varies, as does their duration and intensity. All this reminds us that although these are tropical lowlands, they are on the northern extremities of the tropics and thus subject to some of the seasonality of the continent as well as the tropics. These are not the monotonous, enervating "hearts of darkness" of the novels nor the tropical set-pieces of the textbooks, in which diurnal variations are greater than the annual variations! The weather at Veracruz is not only variable but disorderly, going from extreme to extreme, from heat to chills, from storms to awful calms. There is no equilibrium. Such a place is bad in its temper. The heat and the vapors ponded over it during the rainy season loosen and corrupt the humors; the human body weakens and thins.

Visiting Englishmen ring their own changes on these ideas:

> [Veracruz] is subject to great sickness, and in my time many of the Mariners & officers of the ships did die with those diseases, there accustomed, & especially those that were not used to the countrey, nor knew the danger thereof, but would commonly go in the Sunne in the heat of day, & did eat fruit of the countrey with much disorder, and especially gave themselves to womens company at their first comming: whereupon they were cast into a burning ague, of which few escaped. (Tomson in Hakluyt 1927: 260)

This brief statement includes a whole compendium of old cautions about tropical places. Disorder is to be avoided, that is fundamental. The unaccustomed are especially vulnerable; guidebooks say that to this day, for various reasons. Do not go out in the midday sun! Watch the fruit! Be careful of excesses of any kind and do try not to relax the moral fibers too much. There is now the very serious danger of something worse than the "burning ague."

Hawkes, in 1572, gets so close to insect vectors that one catches one's breath:

> [Veracruz] is inclined to many kent of diseases, by reason of the great heat, and a certeine gnat or flie which they call a musquito, which biteth both men and women in their sleepe; and as soone as they are bitten, incontinently the flesh swelleth as though they had bene bitten with some venimous worme. And this musquito or gnat doth most follow such as are newly come

into the countrey. Many there are that die of this annoyance. (Hawkes in Hakluyt 1927: 280)

What can be done? Chilton observes, in 1568:

For the unwholesomness of [Veracruz] they depart thence six-teene leagues further up within the countrey, to a towne called Xalapa, a very healthfull soile. There is never any woman deliv-ered of childe in this port of Vera Cruz: for so soone as they per-ceive themselves conceived with child, they get them into the countrey, to avoid the perill of the infected aire, although they use every morning to drive thorow the towne above two thou-sand head of cattell, to take away the ill vapours of the earth. (Chilton in Hakluyt 1927: 265)

Waters

The hydrography of Central Veracruz in the late sixteenth century materials is quite disappointing. There is hardly a hint of what has come to dominate our own appreciation of the subject: the strong seasonal fluctuation in through-flowing streams and the yearly dry-ing out and flooding of vast wetland margins. Dry season charac-terizations predominate in the descriptions of the natural environ-ment, as in the opening quote. Little is said of the time of high water. Small wonder; one leaves the lowlands during the wet sea-son if one possibly can. In any case, the "airs" are much more care-fully considered, but then they can readily bring death.

The axis of our region is the river known before contact as the *Huicilapa[n]*, then in the first years of the colonial period as Río de Canoas and by the time of the *Relación* as Río de la Vera Cruz (Hernández Diosdado 1580: 313). Now it is usually called the Río de La Antigua, or more awkwardly, the La Antigua River. The médico admires the river: it has the clearest and healthiest water of any stream in the world. Again, one deduces that he is thinking dry season and not the time of the rains, when it is choked with sediments and has clumps of vegetation and detritus floating on its surface. But then it may not have been so sediment-laden in any season during his time, considering the general retreat of agricul-ture and reforestation consequent on disastrous reduction of In-dian population in the tributary region.

The permanent streams that cross the lowlands to the north and the south of Veracruz are described repeatedly as *caudaloso* (voluminous), as of course they would be during the rainy season. But in the dry season they would still be carrying a substantial flow, when the intermittent streams all around had dried up.

The vicar sets out the fundamentals regarding traffic on this river. It is shallow in its lower reaches and goods must be transshipped at the mouth in order that ships can enter and go up to the town.

Just in front of the town, the vicar also tells us, "ay un estero que nace del rio [a sluggish stream is born of the La Antigua River]" (A. Hernández 1571: 196; my translation). This is an interesting observation and increases one's respect for this man. Although the San Juan runs into the La Antigua most of the year, the rising water of the La Antigua does enter the San Juan system at the beginning of the rainy season and contribute substantially, together with rain within the Basin itself, to rising water levels. And we are told that the *estero* is curiously deep near the junction, deeper, in fact, than the La Antigua. This has become critical to the planning for the modern development of the Basin, as will be seen.

Flying over this confluence during the dry season, one sees the light-toned, sediment-laden water of the La Antigua on the left side, opposite the town, and the darker water of the San Juan, with more humic acid in it, entering on the other side. The vicar verbalizes this in terms of drinkability. The drinking water is obtained from the lighter side.

Wildlife

The authors of the *Relación* and the *Declaración* are asked about indigenous plants and animals, as distinct from those introduced on contact. The replies are profuse and one would like to be able to parlay them into a table of Indian, Spanish, and English common names together with scientific designations. From these one might then move—through labyrinths of reference material—to ecology and a comparison of what once was and the much less that there is now. This is not feasible without specialized input and in any case would be out of phase with the objectives of the respondents. They do not provide quite the systematic surveys that the questioners envisaged; they counter bureaucratese with enthusiasm about how good the fishing, the hunting, and the eating is—their reading on the Basin's favor.

Fish.   The médico points to the infinite quantity and excellent quality of the fish in the fresh waters of Central Veracruz; the lagoons and swamps to the southwest of the town of Veracruz are full of fish—an interesting observation about those water bodies during the dry season. The fish have entered during the wet season, have grown fat there, and now are imprisoned, as was recounted in chapter 2. Little is said of the fish of the sea. (Fishermen and fishmongers of the region note now that freshwater fish are scarce and that most of what is offered for sale is from the sea.)

Certain species are emphasized in the colonial documents. Some can be fairly easily integrated into the current taxonomy, others not. The characterization of fish and other wildlife is highly uneven. *Mojarras,* of the family of the *gerridos,* are caught with nets in the streams and lagoons of the Basin opposite the town. This would seem to be one of the species that will have been involved in the yearly migration from the main stream into the Basin, and the subsequent harvesting of fish in the residual pools, but of all this not a word. Many turtles are obtained in the Basin as well.

The fish of the main through-rivers, the La Antigua and the Jamapa, which runs by Medellín, are most impressive—none more than the *bobo* (*Huro nigricans*):

One of the most notable things in the world, of great use to the people of this city, happens infallibly at the beginning of winter, when the north winds set in, at the end of October and the beginning of November. An incredible number of fish come down the river. They are called *bobos* and are one of the most delicious fish to be had in this kingdom. They come in huge squadrons to spawn at the river's mouth, where the sweet water mixes with the salty, a marvelous, gratifying thing to see. (Hernández Diosdado 1580: 194; my translation)

Other fish of the main streams include the *pampano* (*Citula dorsales,* or *Hynnis hopkinsi*), which, the vicar affirms, is very tasty, as well as the *lisa,* the *robalo,* caught on the coast and in the river, and the *corcobado.* There are a few hints of the more exotic. The vicar mentions sharks and goes on a bit about *roncadores* (croakers), but the main interest is in the eating—one fish better than the other, prepared in all sorts of ways: salted, pickled, boiled, fried, or tucked into *empanadas.*

Wild Land Animals.   The most challenging game animals are the *leones* and the *tigres*—not "lions" and "tigers" so much as pumas or cougars and jaguars, respectively. There are not many of them nor are they very big or ferocious. These are the predators that cattle are exposed to; they mainly go after the calves. One hunts them with dogs, trees them and then shoots them with an harquebus, the common portable fire-arm. Their skins *are* valuable.

The more abundant game animals include rabbits and wild boars, but in first place are the deer—roe deer (*Capreolus capreolus*). They are hunted with dogs and fire arms but do not yield much meat. Cortés's men hunted them; in 1555 Tomson, the English merchant, visited San Juan de Ulúa and found that:

> The Countrey all thereabout is very plaine ground, & a mile from the sea side a great wildernes, with great quantitie of red Deere in the same, so that when the mariners of the ships are disposed, they go up into the wildernes, and do kil of the same, and bring them aboord to eate, for their recreation. (in Hakluyt 1927: 259)

Whether as further objects of the hunt or simply as other interesting animals, the authors both note coyotes, which they try to compare with animals known in Spain. They are like foxes, wolves, even hounds; they howl, eat poultry, and bring down young goats and sheep. There are skunks as well, and squirrels.

There are two curious creatures; the first is the armadillo, the other is one of the most remarkable things, the médico maintains, that one might possibly imagine. It is a bit like a ferret but really different than anything he knows. It carries its young in a pouch, which go out to suckle and climb back in—clearly one of the opossums.

To all this must be added the really bothersome animals—no account of tropical lowlands has ever been complete without them. Poisonous snakes lurk in the vegetation, mosquitoes are the basic plague, and alligators inhabit the rivers and lagoons.

Birds.   Migratory birds are observed to come to Central Veracruz from Florida and cold lands to the north; they include pigeons and hawks. The foliage is enlivened by parrots and parakeets. There are thrushes, like the Spanish starling, and various kinds of herons, which are killed for their feathers.

The birds hunted for food include wild turkeys, found in large flocks—not only in the wild, but also around the houses. Quail can be had and are good to eat, says the vicar. And there are various kinds of ducks.

If we take the food that could be hunted and taken out of the streams, together with the readily available beef and pork, the fruits in the kitchen gardens, and the staples brought down from the uplands, of which more will be said, the eating must indeed have been good in Veracruz. It all assembles itself into a delectable still life, the painting popular in Europe in the sixteenth and seventeenth centuries. Behind it, as some art historians would maintain, is a celebration of God's creation stimulated by the plants and animals discovered in the New World (Wheelock 1989: 18–20).

Plants

The "Memorandum" asks about wild trees, cultivated native trees, introduced trees, native seed grains, and vegetables; about the Mediterranean suite of basics: wheat, barley, wine, and olive oil; and about native herbs and aromatic plants (grasses are not mentioned!) (Cline 1972: 236). The replies are sufficient, when combined with field knowledge, to detail the greenery quite well and to evoke the visual profiles presented in the following figures.

Vegetation around the Town of Veracruz (figure 4.1). Neither of the two main authors sees high forest, but both note thickets of low species. The trees in them are good for little more than poles or firewood. They are generally evergreen and difficult to clear; cut them down and they sprout again with vigor. These thickets are *arcabucos,* and with this the authors put a name to some of the obtrusive, ambiguous greenery that has always crowded around one in the tropical lowlands—neither clearly of human nor of natural origin, certainly not a "climax" community. Their characterization does not sound quite like the *selva baja caducifolia* that Gómez-Pompa indicates for the region (1980: 64), or the chaparral the American soldiers found so infernally difficult to move through in 1847 (Siemens 1990: 133–147). It does seem successional and variable in extent, like the tree growth often described for savannas.

There are many trees within the town; if one had flown over it in 1580 one would have seen roughly what one sees now: streets and main buildings just visible among a substantial concentration of high greenery. Fruit trees and vegetables are listed in the documents; one may deduce circum-residential gardens similar to those of today. Prominent within them are the various introduced citrus trees.

The médico mentions that *Cocos de Guinea* have been brought from Cabo Verde and do very well. He notes that palms of other kinds were already extremely abundant in this area. Instead of clus-

Fig. 4.1. Profile of vegetation around Old Veracruz (La Antigua): (*a*) *arca-bucos;* (*b*) useful trees of many kinds within the town; (*c*) *Cocos de Guinea;* (*d*) maize and sugarcane; (*e*) "wet savanna"; (*f*) "dry savanna"; (*g*) bottom of the La Antigua River; (*h*) lower bottom of the San Juan River; (*j*) settlement.

ters of dates, however, as one might have seen around the Mediterranean, these palms yield clusters of nuts that are commonly called *coyoles.* One thinks of both the coconut palms that can be seen just southeast of the town today and the fine stands of *Roystonea dunlapiana,* with their bunches of nuts, in the pastures of the Basin, as well as the many other palms of the region that have not yet had the botanical attention they merit.

Both the vicar and the médico have a good deal to say about the wood that is brought into Veracruz from its surroundings. There are many wild trees in what is perceived as the ambit Veracruz that are useful for the construction of houses—and the repair of ships, a lasting concern along these coasts throughout the colonial period (Siemens and Brinckmann 1976).

In first place is *cedro,* probably referring to *Cedrela mexicana,* which is known now as *cedro rojo* and is a characteristic species of the *bosque tropical perennifolio* community (Rzedowski 1978: 163). The trunks are thick, long, and straight, the wood is easily sawed but durable; from it, the médico maintains, are made the best beams and decking in the world. It is found on Indian land in the vicinity of Medellín—a marker of the higher, more fully evergreen tropical forest that sets in as one moves out of the semiarid climatic enclave prevailing over the San Juan Basin.

There are various other good structural woods, probably also found to the south, as well as in the canyons to the west, in the gallery forests and indeed any higher ground within the Basin, particularly the eroded remnants of Tertiary sediments. The mulberry is named—excellent for ships and houses. There is a tree for which the médico finds only the Indian name *tenezquauitl* (Indian words are seldom used in our sources!). It is a white wood, useful for the building of carts. A tree much admired is the *mamey (Mammea americana);* its trunk is large and its wood very strong, it is used in the framework of the largest ships of the time, and is brought to Spain for the beams of wine and olive presses.

The fruits available in the coastal region have fascinated outside observers from the time of first contact. Neither indigenous nor introduced fruits are actually grown much in and around Veracruz itself. Pomegranates, figs, and grapes had been introduced before our documents were written and had done well for a few years, but then their roots had decayed because of the heat and humidity, so no one had bothered further with them. Only the citrus fruits are abundant in the kitchen gardens within Veracruz, and on the lands around the town. There are no plantations—for lack of Indians to tend them. Excellent fruit is brought in from surrounding Indian communities, from Cempoala, Medellín, and Rinconada. This is what gives exuberance to the markets and does much, as the médico points out, to mitigate the desire for the things of Spain.

There are links upslope into the temperate zone above Jalapa. Pine is brought down (in the form of shakes, for roofing), along with apples, pears, and grapes.

As regarding medicinal herbs and aromatic plants, the médico observes, there are no Indians from whom one might gather a proper response. Various admirably effective purges are commonly known. (They rank with snakes and with heat and humidity as stock elements in the characterization of tropical lowlands from Contact to modern times). The vicar includes the most common one, *zarzaparrilla*, a creeping epiphyte. The Indians of Medellín sell it to dealers, and to seamen when the fleet is in.

There is little agriculture around Veracruz; what there is is clearly relegated in the descriptions. Some beans and maize are grown, but these are food for Indians and blacks, horses and mules. Maize was once grown in abundance here and the land is excellent for it. It is not difficult to deduce where: on the levee slopes opposite the town, which is land now used for commercial cropping. There are quite likely to have been wet season fields on the older dunes as well and on the hill land to the west. Alba González Jácome traces early-sixteenth-century subsistence production along the road from Veracruz into the interior (1997: 96). The wheat required by the Christian townspeople and the ships crews comes from the Mesa Central.

The médico is definitely not a man of the land, but he sees good, heavy soil, with what seem to him to be all the things needed to cultivate wheat. He speculates on why it is not done. Perhaps it is because it rains here at the time of year when in Spain one would harvest—some other timing had not apparently occurred to anyone; perhaps it's because everyone is busy with other things.

Grasslands on Terra Firma.   It is difficult to proceed with the profiles without addressing the ambiguities of the term savanna in this context. An Arawak Indian word for grassland, it came to be used on the mainland by the Spanish conquistadores and then much more widely down to our time. It is variously spelled, repeatedly defined and redefined, but can never really be simplified because what it represents remains complex.

Savanna is considered to be a variable association of grasses and trees controlled in very general terms by climate. It occurs in tropical areas with strong seasonality; that is, a dry season of at least several months duration (Harris 1980: 3–5). It can be seen on various soil types and is botanically highly diverse—indigenous species supplemented by introduced plants (Jordan 1993: 10).

There is a basic ambiguity with respect to its origin:

> We now have much evidence to support the statement that savannas did exist in tropical America before the arrival of the Europeans, before the development of aboriginal cultures, and most probably before the arrival of *Homo sapiens*—even before the birth of this species on the planet. (Sarmiento and Monasterio 1975: 235)

However, even though the origin of the savanna cannot be considered as anthropogenic, it cannot be seen as entirely natural either. The location and extent of savannas is likely to have been affected by human intervention, particularly the use of fire, in places that were attractive and in which natural factors favored the change (Hammond in Harris 1980: 98; Rzedowski 1978: 235).

Gómez-Pompa has brought this line of thought to bear on the grasslands of Central Veracruz (1980: 70–74). Some savannas of the Gulf Lowlands, such as those described for the swampy bottomlands of the Basin, seem clearly of natural origin; that is, they are edaphically controlled. Elsewhere, as in the sloping hill land to the west of the Basin, the characteristic tropical forest of the region, *selva baja caducifolia*, has been given a *sabanoid* character by periodic burning.

Cattle themselves alter grasslands as well. Grazing stimulates plant growth; the animals fertilize with their manure and carcasses, they disperse seeds and select among plants, changing the floristic composition of their pasture (Jordan 1993: 10).

Verdor perpetuo (figure 4.2).   The médico, in his opening effusion, sees subdued, well-watered terrain around Veracruz, especially

Fig. 4.2. Profile of vegetation in the western San Juan Basin: (*a*) San Juan River; (*b*) levees with gallery forest; (*c*) backswamp with hydrophytes; (*d*) "wet savanna"; (*e*) savanna *brava*; (*f*) corrals and milking stations on "islands" of tertiary sediments within the wetlands; (*g*) ranch headquarters or other settlement on terra firma; (*h*) "dry savanna" with *arcabucos*.

south of the town, and freshness, perpetual green, beautiful pastures ("hermosisimas dehesas y sabanas"; Acuña 1985: 312–314). This is a dry-season view from a height, with breezes blowing from the ocean. Later come real heat and humidity, but by then he is probably in Jalapa. The vicar looks south and sees pastures too, with swamps and high, rank savanna ("toda empradescida y llena de ciénagas de cauana brava y alta"; A. Hernández 1571: 194).

Two words here are quite interesting: *empradescida*, from *empradizar*, and *dehesas*, from *dehesar*, both meaning to convert into pasture (Williams 1955). In both cases the resulting grassland has been induced and is distinguished from savanna. This grass is shorter than the vicar's savanna *brava*; but both would be covered by the term *wet savanna*. The induced grass is lusher growth than one would see in the grasslands of the hill country to the west, the *dry savanna*. This indication of change from forest to pasture seems to link up with information in the basic documents about a considerable number of slaves on the Ruíz property in the middle of the Basin. Ranching does not require much manpower *per se*, but clearing does. It also links up with the evidence for historic forest clearing in the upper strata of our own excavations in the San Juan Basin (Siemens et al. 1988). The key vegetational impact of the Spanish design in these parts was the retreat of tree growth and the tall hard grasses in areas subject to inundation. But these would surely advance again rapidly if the pressure of ranching were reduced. The vegetation probably pulsed repeatedly in this sense in the colonial period and the nineteenth century.

The vicar is especially taken with savanna *brava*. This grass is as tall as a man. The stalk is a substantial cane; its leaves are long and thin—useful for thatch. It is edible for the cattle only when it is kept low by fire.

A group of us came on a brake of such grass one day during a search in a wetland margin near La Antigua for some remains of ancient wetland agriculture that we had detected from the air. The grass was a solid mass, well over our heads, with canes several centimeters thick and the bases of the plants in standing water. There was no penetrating it, no way at all to find the slight variations in the surface represented by the linear patterning evident from the air, short of burning it all down. We found such grasses in many other places in the lowlands, in swamps and immediately around open lagoons or along streams.

On the Hills and in the Valleys to the West (figure 4.3). The vicar compares the short grass after a fire in the humid lowlands to the less luxuriant grass that grows westward of the swamps in the "cerros y partes no tan humidas [hills and areas not so humid]" (A. Hernández 1571: 195; my translation), presumably during the rainy season. Moreover, "en partes ay vegas descubiertas una legua y más [in places there are cleared riverside plains for a league or more]" (A. Hernández 1571: 195; my translation). These plains were used for pasture and for sugarcane (González Jácome 1997: 97). Cane requires not only humid soils such as those found in the river-valley bottoms but also good drainage, and hence is kept above the zone of inundation.

Immediately after this the vicar names, "*Maguey y arboles de algodon* [maguey and tree cotton]" (A. Hernández 1571: 195; my translation). They are likely to have been found on the piedmont surface. This is also where the remains of the Prehispanic stone lines are found. The *selva baja caducifolia* shown on the graphic profile is deduced from what is characteristic now (Gómez-Pompa 1980: 64), and the *estancia* from both characteristic locations seen on sixteenth-century maps and common locations of ranches at the present time. Intriguing place-names on modern topographic maps raise the possibility that one or another of the valleys were routes of transhumance.

Spanish Settlement

Much of the environmental information purveyed in the two main documents can be arrayed on profiles. The new mesh of communities and transportation routes invites a planimetric representation. And the *Declaración* and the *Relación*, with the land-litigation maps alongside and our own field experience in the background, suggest certain areal dispositions.

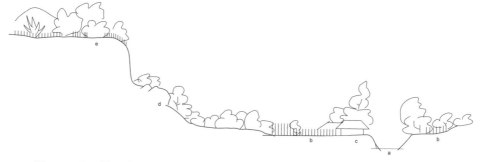

Fig. 4.3. Profile of vegetation on one side of a wide *barranca,* in the hill land to the west of the San Juan Basin: (*a*) one or another of the streams flowing eastward in Central Veracruz; (*b*) pasture and sugarcane on the *vegas* along the stream; (*c*) *estancias;* (*d*) rock falls and rough slopes below the cliffs; (*e*) surface of the piedmont with a mosaic of *selva baja caducifolia* and "dry savanna," interspersed with *maguey* and tree cotton.

Evoked Cartography

"Evoked cartography" employs free strokes, as a researcher might respond in a notebook to observations on directions and distances made by informants in the field. The maps that accompanied land litigation were clearly drawn in this way (e.g., figure 4.4).

Today in La Antigua, the "Veracruz vieja" of early colonial times, one can meet one's informants at a table in an open-air restaurant near the landing of the dugout "ferry." It is not difficult to imagine a discussion in 1571 or 1580 with voluble informants in the shade on that same riverbank. Vegetation will have blocked any actual view southward then as it does now, but locational affirmations will have been made and countered with a good deal of pointing in various directions, and perhaps some actual sketching.

Those who drew had no pencils or erasers. Rough drafts may well have been done, even so, but one gets the strong impression that the draftsman took a deep breath, dipped his quill or his pen, and just started in—like as not from the bottom of a sheet of paper in this case, putting south at the top of the page. He would be committed further and further with each stroke. He could make minor corrections, and the maps have many examples, but the process was incremental for better or worse. He could make adjustments in direction and scale, particularly in scale, as he went along, making judgments as to the space still left and the relative importance of this or that feature, perhaps responding to the remarks of a client looking on or even a company of kibitzers.

Fig. 4.4. An early colonial land-litigation map. South (*sur*) is at the top and north (*norte*) at the bottom. The wavy line at the bottom is the La Antigua River and the one at the top is the Medellín River. The critical element of this map is indicated by the tree symbol—the *sitio* or *estancia* that is being solicited (Archivo General de la Nación [AGN], *Tierras*, Vol. 2764, Exp. 15, f. 188 [1592]).

The locational information in the *Declaración* and the *Relación* is written with a southward orientation; it seemed useful to let the principal map that emerges from this analysis conform to that (figure 4.5). Acuña's analysis comes to grief because he has not recognized this, as we have seen. At least two pre-publication reviewers of the manuscript for this book protested. It is just not done; it is confusing. School courses in geography have conditioned all of us to certain conventions and reference systems; computerized drafting has made certain procedures common. Aside from the orientation, figure 4.5 adheres to much of this. The first drawing for the figure was redrafted, digitized, and edited, as a matter of course. The resulting map is thus a hybrid. It sets out much of the sort of information that a *pintura* was meant to convey but can be more disciplined. However, it is far from a direct placement of the locational material on a topographic base (cf. Siemens and Brinckmann 1976: 269). Relative location in this context is more accurate than actual location; the whole is quite schematic. It allows the material to express itself at least partly in its own sixteenth-century terms, but for our purposes: a reenactment of the representation of the early Spanish imprint on this landscape.

The map reflects the central fact that a view from the Veracruz of the time tended to be a view southward. That is where the mouth of the river was, a little farther on the main anchorage of incoming ships at San Juan de Ulúa, and still farther on the main rival town, Medellín. To the south, too, was that vague, swampy expanse with the many *estancias.*

Repeated cross reference between the primary materials and topographic maps generated a scale of five kilometres to the league. No specific definition of the league could be used, such as, "The league of the country measures, in terrestrial longitude, 5000 *varas* or 4,190 metres" (Santamaría 1978: 658; my translation). It had to be an approximation roughly in accord with the very approximate scale that would exist in the minds of a group of informants in the shade on a riverbank.

The médico sets the perimeter of the region in the statement quoted at the outset: "Within a radius of seven leagues are normally grazed more than 150,000 head of cattle and mules" (Hernández Diosdado 1580: 314; my translation). The seven leagues are only very sparsely covered northward; the information thins out quickly westward, too, into the hill land; it is lopped off eastward by the sea. Most of the material in all the sources combined, presented in very generalized form, relates to the space between Veracruz and its weak opposite pole, Medellín, something over five leagues to the south.

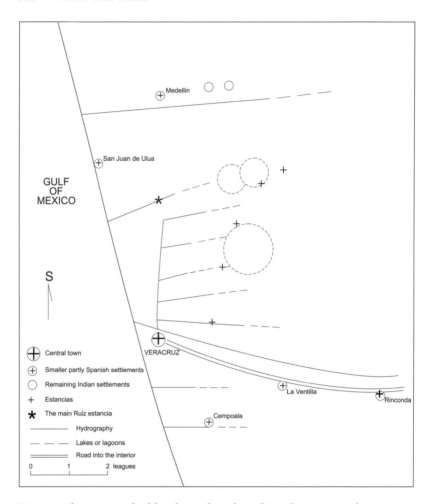

Fig. 4.5. The map evoked by the early colonial *Declaración, Relación,* and related materials.

The médico's seven leagues extend well south of the San Juan Basin, as do six of the seven land-claim maps. The Basin is really only a segment of a region that continues southward, physically and in terms of occupance and use. There are many wetlands and lagoons to the south, some of which we know are extensively patterned. One can therefore envisage other favored places in that direction, but not as accessible in the documents as the San Juan Basin.

There is a stridency in the areal references of our texts, in the seven land-claim maps, and in our own reenactment, reflecting a "political unconscious"—the hidden processes that legitimized and reinforced colonial power (Harley 1992). Land is being appropriated with strong lines and bold circles, the pen firmly grasped, and with confident statements—this or that is what's wanted.

The Indians have been left, mostly, in blank spaces or on the margins. In these early colonial maps, in those that will be introduced in chapter 5, and generally in much of the Hispanic cartography of the Americas it is indeed apparent that, "Conquering states impose a silence on minority or subject populations through their manipulation of place-names. Whole strata of ethnic identity are swept from the map in what amounts to acts of cultural genocide" (Harley 1989: 66).

The Remaining Indians

Depopulation is far advanced by the time the *Declaración* and the *Relación* are written. However one calculates it, the decline in the surroundings of Veracruz reaches at least 90 percent by 1580 (Siemens 1995a: 208). Many of the Indian communities are gone without a trace, others have been amalgamated in order to conserve a vital minimum. "Thinking of what these communities once were, it is sad to see how they have been diminished" (Hernández Diosdado 1580: 188; my translation).There are Indians still at Cempoala, to the north, at Rinconada to the west, at Xamlolulco (Xalcomulco?) just over the river, due west of Veracruz, and farther away in Xalapa, and then to the south, along and just beyond the Xamapa River. The names of these last communities, the vicar notes, are still known among Indians but not among Spaniards (A. Hernández 1571: 194).

The questions asked of the médico in this regard were brusque and those asked of the vicar were probably similar:

State whether there are many or few Indians; whether there have been more or fewer in former times, and any reasons for this that may be known; whether or not they are presently settled in regular and permanent towns; the degree and quality of their intelligence, inclinations, and way of life; and whether there are different languages throughout the province or some general language that they all speak. (Cline 1972: 234)

The answers of the médico are more fulsome and more thoughtful than those of the vicar, and a little less tired, but both depart from assumptions of superiority and sound condescending from where we are.

The Indians are few and poor, they live miserably and show little ingenuity or skill. They are improvident, caring only to gain what they need for the year. But they do pay tribute, which goes to pay the salary of the alcalde mayor.

As has been noted, the vicar, or whoever transcribed his answers, repeatedly strings laconic observations together into verbal collages that entangle the modern reader. One cannot resist reading between and around the words, and judging the Spanish colonial "project." So here:

> They pay tribute in money and maize, which they cultivate.
> They go clothed; all have been indoctrinated. They grow melons, cucumbers and a great deal of fruit. They fish in the river called Cempoala, which is a substantial river. (A. Hernández 1571: 193; my translation)

There is more in this vein: Indian men are excessively given to drink. And because they are mentally weak, a little affects them a lot. One can get access to the *cedros* near Medellín by providing the chiefs with jugs of wine. The Indians are given their masses and sacraments by itinerant priests but cannot be considered good Christians. There are still always intimations of idolatry.

Our sources say nothing of agriculture in the wetlands. Integrated exploitation of wetlands and neighboring hill lands may have been going on, it is not precluded by what is said about the location of still-populated and depopulated Indian settlements in the region or about their life and work, but there is no explicit reference to it either.

The médico indicates that he would like to be able to comply with Item 26 of the "Instruction" but cannot really. It inquires about the "Herbs or aromatic plants with which the Indians cure themselves, and their medicinal or poisonous properties" (Cline: 236). There are no Indians in the immediate vicinity of the town to ask. (One does wonder sometimes what banter surrounded the compliance with this list of questions. They were easy for the bureaucrats in Spain to ask, not so easy to answer.)

One is taken aback that in 1577 the administrators in Spain still needed to ask about the causes of the decline of the Indian population, and that it was answered the way it was. The médico maintains that it is still a puzzle. No one in these parts can think of any other reason than the climate and the plague of mosquitoes. However, he also points out, there is some history relevant here. The language of the Indians of the region is the language of Central Mexico. The lowlanders were originally forced down here by other Indians, their oppressors knowing that people do not conserve well in this climate. Well, it had its effect.

The Town of Veracruz

Veracruz wanders the coastal lowlands during New Spain's first century—enough to confuse the casual reader of histories and even the odd historian, such as Acuña. The first Veracruz was an encampment on the sands opposite the island of San Juan de Ulúa, where the present city would be built. The second was established soon thereafter near Quiahuixtlan, thirty-five or forty kilometres north of the mouth of the La Antigua River. The third was established on that river in 1525, at the site of the present town of La Antigua. The fourth began to take shape late in the last decade of the sixteenth century on the sands opposite San Juan de Ulúa where Veracruz is now (Trens 1947: 30–31, 114–115; Ramírez Cabañaz 1943: 11–13; Rees 1976: 29–30).

Various other moves were proposed, but came to nothing. After the disastrous flood of 1552, the alcalde mayor proposed a move westward up the main road into the vicinity of Rinconada. Others suggested a move even farther upslope, to the *venta* of Lencero, just below Jalapa (Ramírez Cabañaz 1943: 10–11).

Medellín, the antipode of Veracruz in the *Declaración* and the *Relación*, hoped for a time that it would become the main port on the Gulf. Both were on rivers that provided shelter for shipping, but the latter was easier to sail into. Medellín was surrounded by swamps and could hardly be approached by land in the rainy season; Veracruz was better connected into the interior. So, for reasons of trade and not health, as the médico has it, the Spaniards already settled in Medellín moved to Veracruz from 1525 onward (A. Hernández 1571: 193, n. 1; Hernández Diosdado 1580: 310; Rees 1976: 29). Medellín and its surroundings became part of that shadowy world of Indian communities about which the Spaniards knew and cared little, except that it was the region from which the impressive structural woods and the many delicious native fruits were obtained.

Veracruz persists but also has retained a very bad reputation. It is a sick, vexatious place, a tomb of the living. Months are required to unload and reload ships. Meanwhile merchandise spoils or is robbed and crews are decimated by disease (Trens 1947: 128–130,

156–160). By the time the *Declaración* and the *Relación* are penned, conditions seem to have improved somewhat, or perhaps the authors are just putting the best face on things.

In any case, the médico makes Veracruz into the first stone of Spain's grand edifice in New Spain (Hernández Diosdado 1580: 310). More vital images suggest themselves: Veracruz is the navel of New Spain, the road into the interior is the colony's major artery. This road has no Prehispanic precedent; the Aztec route into the lowlands went via Tehuacan southeastward toward Tabasco (Rees 1976: 18). The colonial imperative was to transit the lowlands as expeditiously as possible. This reorientation of transportation focused on Veracruz represents the colonial succession as clearly as any other Postcontact change in the lowland landscape.

Veracruz itself is in, but not of, the lowlands. It draws on the resources of the surroundings to a limited extent: meat and other foods (except staples) for the consumption of residents and transients, pasture for draft and pack animals. Little of the merchandise imported is for the immediate retail market; the merchants' interests are much farther flung.

Veracruz can be approached easily on the prevailing breezes; it provides an anchorage safe from the *nortes,* not far inland from the mouth of the river. But the ships can only come in riding high; the bar at the mouth makes transshipment necessary, and that can last months. Often ships anchor behind the rudimentary protective wall on the island of San Juan de Ulúa, five leagues to the southeast, and have their goods shipped first up the coast and then along the river to Veracruz, which of course takes much longer still. The road into the interior departs from Veracruz along the grain of the landscape; that is, roughly parallel to the *barrancas* that striate the piedmont. There are no serious streams to ford and a reasonably gradual slope to ascend—which is more than can be said for the route that leaves from Medellín. This, then, is the vital route of entry into New Spain.

Veracruz sits on the north side of the river, on the first bit of solid ground west of the dunes (figure 4.1). It is opposite the mouth of the stream that is identified only as a large *estero,* which we have come to know as the San Juan River. Within its basin are many *estancias.* There is good fishing there, especially late in the dry season, but from that direction too come the airs that help to make Veracruz unhealthy. The bank on which the town was built is subject to flooding in years with extraordinary rain—it is more vulnerable than the opposite bank, as we have seen.

According to the vicar there are two hundred *vecinos* in Veracruz.

This term has come to be interpreted as including the owner of a house and those that live under his roof: his wife, single children, relatives, servants, and slaves (Gerhard 1986: 27). Thus one might estimate that there are between 2000 and 2500 people living in the town. The médico notes fewer *vecinos*, but then his count is a decade later. Sluyter has used a lower conversion rate of *vecinos* to total Spanish population and also shows the population of the town in decline over this period—it had several thousand people in the 1560s and then dropped to several hundred toward the end of the century (Sluyter 1995: 160, 485).

There are no Indians or mestizos in the vicar's Veracruz, but more than six hundred male and female black slaves. There are two hundred additional slaves out on the main estancia of Hernán Ruíz. The town also has a few freed blacks and some *mulatos*. The population swells with transients of the sea and the road when the fleet comes in; it drops off during the summer, when as many Veracruzanos as are able evidently go upslope to live in Jalapa. More than five hundred people, the vicar notes, are practicing Christians.

The houses of Veracruz are of adobe or brick, divided into rooms by cloth hangings and roofed with tiles or shakes. The roofing is brought from the mountains, as are pine boards and beams; *cedro* is brought from Medellín. The roofs of Veracruz were once all of thatch, but there was a fire that damaged forty or fifty houses in 1565 or 1566 and since then thatch has been prohibited.

Religion and charity are prominently represented in the structure of the town. There is a parish church, of stone and brick. There is a Franciscan monastery and a Jesuit one under construction, both financed entirely from the offerings of the faithful, as well as a large new church that honors Nuestra Señora de Consolación. The vicar mentions several hospitals; the médico notes only one, but stresses its importance for the sick coming in on ships. In fact, the ships contribute to its support. Here then are the vector-borne diseases of the lowlands striking particularly hard at the outsiders coming in without having developed immunity through mild infections in childhood. The vicar remarks that alms are given liberally in this town; he tells of one Sunday when four thousand pesos were given. One can imagine some reasons! Many a traveler will have been very grateful to have survived a *norte* but very concerned, too, about survival while in Veracruz and safety on the journey into the interior. It was an exciting place.

There are other facilities: a *casa de contratación* to regulate and tax the trade, and a *casa de cabildo*, to serve the administration of the town. There is no fort, but a garrison of four regiments of soldiers. There are various meat shops, but no whorehouse, the vicar says, all in one line! (He was almost certainly turning a blind eye.) Just out of town, there is a landing on the river with a storage shed and a crane.

In and immediately around the town there is the profusion of greenery already described. Woodcutters go out into it for fuel and building materials; carpenters and cartwrights ply their trade in the town. Agriculture is sparse: some maize, but no wheat. No one has really tried it in this region. However all this is not a problem since the mule and cart trains coming down from the uplands would otherwise go empty, so it is relatively easy to amply supply the town and at the same time victual the ships.

The main work in Veracruz is trade. This, in turn, requires shipping and cartage, in which slaves are used. People get rich through trade and the renting out of slaves, as well as the renting of houses, warehouses, boats, and wagons when the ships come in. And not to be forgotten in the economy of the town, as well as the region, are the *estancias*. Their owners have houses and trading establishments in town.

The vicar points out that there are ten or twelve men with a capital of more than 20,000 pesos. The man of substance in these parts is Hernán Ruíz de Córdoba. He owns two hundred black slaves and twenty *estancias*; these are clearly the important measures.

Robert Tomson, an English merchant, arrived at San Juan de Ulúa on April 16, 1555. His party had lost most of their possessions during a strong *norte* in the Gulf of Mexico. They made their way northward along the beach and were amazed at the large uprooted trees that had come ashore—from Florida during the same storm, they were told. In Veracruz they were graciously received into the house of a very rich man, Gonzalo Ruíz de Córdova, an ancestor of Hernán. They were kept there a month, outfitted anew, and then given money and the means of transportation up to Mexico (Hakluyt 1927: 253).

Sluyter calls the Ruíz de Córdova family the most successful early land grabbers of these parts; he reviews the history of early colonial land tenure in Central Veracruz much better than can be attempted here (1995: 295–304). This puts chapter and verse to the process of durable legal and ecological succession referred to repeatedly in this analysis.

The trading establishments of Veracruz are not described directly. Trens cites a description of goods coming ashore in 1597, which helps us to visualize what will have been stored and traded in Veracruz: wine, in the first place, then syrup, vinegar, oil, paper, cloth, velvet, spices, books, and iron fittings and other hardware (Trens 1947: 212). It is difficult to avoid recollection of the rudimentary, multifunctional trading houses found by this author in the tenuously connected outlying settlements along the settlement frontiers of the Gulf Lowlands in the 1960s. Storage, conveyance, labor exchange, and money-lending, as well as buying and selling of a wide range of goods at retail and wholesale levels—not forgetting the sale of spirits—all this and more went on in these interesting establishments and may, perhaps, be projected into the early colonial lowlands. Looking into such a place in sixteenth-century Veracruz one will have seen substantial buildings and courtyards, a pile-up of goods, animals, and men coming and going, foremen and slaves, noise and curious smells, presided over by the factor or his son.

John Chilton arrived in Veracruz in 1568 (Hakluyt 1927: 265). He counted four hundred merchants and noted that they stayed in town from August to April—the length of time it took to unload and load! After this they all went up to Jalapa to avoid the rains and the most dangerous, uncomfortable time in Veracruz. No women are delivered of children in this town, he maintains; they went up to Jalapa to have their children. This fleeing to Jalapa became one of the "facts" that was widely known about this coast; it is echoed in various travelers' accounts in the nineteenth century (Siemens 1990), although neither the vicar nor the médico mention it. It must have been something that only a minority could afford; most residents will have had to stay all year round.

San Juan de Ulúa

During the last half of the sixteenth century there is repeated word of a small settlement developing five leagues to the south of Veracruz, on the island of San Juan de Ulúa and the adjacent mainland. This would be the fourth Veracruz, where the first had been and where the port is today.

In 1555 Robert Tomson found a house and a chapel on the low island of San Juan de Ulúa, as well as the very important stout wall on the northwest-southeast-trending landward side. The wall, often damaged during storms, needed to be kept up by twenty of the king's "great mightie Negroes" (Hakluyt 1927:259). Ships would fasten bow-wise to rings in the wall and throw anchors aft in order to ride out the *nortes*. This wall would remain the main haven along this coast for many years to come.

John Chilton arrived in 1568 and found that the King had 50 soldiers and their captains on the island, as well as 150 Negroes—to maintain the wall, which now had two bulwarks, and to help fasten the ships (Hakluyt 1927: 265). The slaves will have been quarrying coral, which became the principal material for construction on the island and the mainland—a stone that was light, relatively easy to cut, and absorbed the shock of a cannon ball so that a wall made of it would not easily crumble.

The vicar reports that the slaves live in twenty houses on stilts; the island is awash during high seas, such as those that accompany the *nortes*. He also notes a church and a hospital.

Henry Hawkes reports in 1572 that the Spaniards are constructing a fort (Hakluyt 1927: 279). It is clearly well underway by the time the médico is preparing his *Relación*.

Fray Alonso Ponce landed at San Juan de Ulúa in 1584 (Pasquel 1979: 225–236). He was detained there by the weather and probably worked on his journal for something to do. We owe various detailed descriptions of the fort, the islands around it, and indeed the town over on the mainland, to such detentions down through the years. Ponce is best when he tells us what it was like during a *norte:* wind, sand, and spray coming over the walls. One could not cross the plaza of the fort, to say nothing of crossing over to the mainland; eyes were blinded and clothes ruined.

There is not much on the mainland. Tomson found it "very plaine ground, & a mile from the sea side a great wilderness, with great quantitie of red Deere in the same" (Hakluyt 1927: 259). These served for recreation and food. The map that accompanies the 1580 *Relación* de Tlacotalpa shows two *ventas*, the Venta de Ramírez and Venta de Buitron (Acuña 1985: 288–289). Our authors note that the prospect is rather cheerless, what with the sand dunes and all, as visitors to these shores were to remark again and again in the ensuing centuries. They also note that water for the ships can be obtained from a lagoon not far from shore. Quite curious this water, it is not very appetizing when first obtained but improves in the casks on board, with time! Firewood can be obtained from the wrecks strewn on the shore.

Veracruz was formally moved to its fourth location, opposite the fort, in 1599. There were some fine rearguard arguments against the move. The third Veracruz had much more solid storage facilities, especially for the many tons of wine that were brought from Spain. The fourth site was a Libyan desert; for water, there was only a swamp, which cattle entered and fouled (Rees 1976: 29–31). This will have been the water that improved in the casks! To no avail; the town did move.

In 1609 Fray Alonso Mota y Escobar, the bishop of Tlaxcala, came to the third Veracruz, which is now La Antigua, while on a journey through his diocese (González Jácome 1987: 55). In his account, and probably in common parlance by that time, the town had become "Veracruz vieja." The houses were deserted and in ruins, which he found unfortunate because they had been such good houses. The parish church was crumbling. There were only eight Spaniards in the town. Various officials were still in place; there was a priest, whom he had to replace in response to popular demand. Most of the townspeople were black—freed slaves. The bishop preached to them in their own church, confirmed sixty-seven children, and consoled them all, as he says, with love.

In 1697, about a century later, Juan Francisco Gemelli Carreri arrived at "Veracruz nueva." He found it sad and boring, so he went hunting in the San Juan Basin. He crossed the La Antigua River by canoe and visited "Veracruz vieja." He saw only temporary fishermen's shelters of thatch surrounded by cane fences and noted that the mosquitoes were terrible (1983: 249).

Ranching

The médico gives us 150,000 head of cattle within a radius of seven leagues, but otherwise is not very helpful about ranching. Sluyter argues that even this is not reliable; this fine rounded figure that has launched many reflections on ranching in coastal New Spain pertains to a larger region and, in addition, indicates grazing on the *estancias* far in excess of the numbers stipulated by the grants (Sluyter 1995: 279–281). That disturbs inherited wisdom in several respects.

The vicar, who knew the everyday Veracruz, reviews the introduced domesticates. He begins with chickens and pigeons, many of which are feral, and goes on to cows and bulls, many of which are feral too, and donkeys, mules, and horses, some fine, some not so fine. There are many pigs; their meat is good eaten fresh; that is, dry-cured, he seems to be saying. There are goats, but not many. Butter, milk, and a great deal of cheese is made from the milk of cows and goats. There are no sheep, he informs us, except those that come down into this region from the highlands each year to graze in the harvested fields and, curiously, those that must be delivered for local consumption. Apparently the tithes due to the church in Veracruz in the sixteenth century were payable in pigs and sheep (González Jácome 1997: 99); this may be what the vicar is referring to. In any case, veal is better eating he tells us—in fact, it is the best of meats.

The lowlands emptied of Indians are being taken up for ranching by means of the land grants called *estancias*. Much very useful work has been done on this and other aspects of ranching in New Spain recently (e.g., Butzer 1988; Butzer and Butzer 1995; Doolittle 1987; Jordan 1989, 1993); it remains here only to test and elaborate on various points. *Estancia* is virtually synonymous in our materials with *sitio* and *hato*. Legally it was considered a unit of land of a certain size: 3000 *pasos* square (1 *paso* = ca. 140 centimeters) for cattle and horses, 2000 *pasos* square for sheep (Butzer and Butzer 1995: 155). On early colonial maps the *estancia* was sometimes conceived of in circular terms, at other times it is represented only by a sketch or a symbol for its central facilities. Sluyter has showed that the *estancias* can be mapped with areally specific squares—by detailed epochs in the granting process (1995: figs. 5.8–5.16, for example). The analysis that ensues and the critique of preceding work in which attempts have been made to bring the documents down to earth is quite absorbing.

Ganado Menor

An important distinction is frequently made in the literature: there are *ganado mayor*, that is cattle and horses, and *ganado menor*, which are sheep and goats. Something like twenty-five sheep *estancias* and eighty-five cattle *estancias* were granted within our region between 1550 and 1615 (Butzer and Butzer 1992: 9). It may be imagined that this was not the whole story of land acquisition: there were undoubtedly unauthorized land seizures (Jordan 1993: 96); also, some of the grants may not have been actually taken up. Be that as it may, the sheep *estancias* were evidently assigned explicitly for winter grazing. This elaborates on what the vicar and the médico seem to be saying; the sheep were seasonally in the lowlands but not headquartered there.

Butzer and Butzer have found that "By 1596 at least 750,000 sheep (and probably twice that many) annually streamed across the mountains from Jalapa to Orizaba, to winter on assigned *estancias* in the foothills or near the coast" (1995: 163). The médico remarks that the yearly invasion involves innumerable sheep; a substantial movement is thus going on along routes to the north as well.

One may deduce movements of hundreds of kilometres, from the rainfed pastures of the uplands eastward with the onset of the dry season, which is also, as it happens in upland Central Mexico, the onset of winter. The flocks proceed downslope over the remains of harvests, through various microenvironments, into the just-burned wetland savannas of the lowlands and then back up when the rains come and the hill lands, including the savannas, have greened.

Many aspects of this transhumance remain to be examined: Did the flocks move down the flat surfaces of the piedmont or through the canyons that striate this landscape, or both? Are there remains of walkways (*cañadas*) yet to be found? How much can be read from the place-names? (Paso de Ovejas on the Veracruz–Jalapa road is the most obvious example.) Were only particular species involved? Where did shearing and lambing take place, or the boiling down of mutton for the fat?

A tropical lowland climate is stressful for sheep—and goats. Sartorius would clarify this in the nineteenth century: humid ground affects the hooves and gives the animals chilblains, spines and seeds from tropical brush and ground cover ruin the wool; in the wet season it is particularly difficult for the animals, what with ticks and the worms that enter any lesion (1870: 180). Melgarejo would affirm, later still, that Veracruz never was "tierra de ganados menores" (1981: 62). How then to explain the references to transhumant sheep and the *mercedes* for sheep *estancias?* The forays into the lowlands do seem to have been undertaken when temperatures were mildest and into pastures where the rank growth had been newly reduced by fire, but were there other adaptations? And what was the subsequent history of the phenomenon?

Ganado Mayor

Two affirmations occur repeatedly in any discussion of early colonial occupance and use of land in the Gulf Lowlands of Mexico: The one is that "[Bovine] Cattle spread like the waves of a rising tide" (Chevalier 1963: 94), and the other is that most of these cattle were feral (Parsons 1989: 317; West 1969: 118; Dampier 1906, 2: 157–160, 179).

The locus of husbandry, such as it is, is the *estancia*. It appears on the landscape as "a seasonal headquarters for a particular (but mainly implicit) number of animals" (Butzer and Butzer 1995: 155). Our texts and ancillary maps offer some clues to their nature and location. One may deduce streamside locations for various of the larger *estancias*. The facilities could have been on the levee tops or, more likely, on higher ground near one of the many streams flowing into the bottomlands.

Some *estancias* are marked as *poblado*—populated. Some appear as solid, arcaded buildings on an eminence and still others simply as conical heights. From maps out of areas to the north and south of the Basin it is apparent that *estancias* varied from fairly substantial ranch headquarters with permanent housing for the hands, to outlying thatched shelters (Archivo General de la Nación (AGN), Ramo Tierras, Vol. 2680, exp. 11, f 155 [1585]; ibid., Vol. 2686, Exp. 3, f 110[1587]; ibid., Vol. 2686, Exp. 5, f 134 [1589]).

The outlying *estancia* of the sixteenth century is probably approximated today by facilities that ranchers call *ordeñas:* shelters, corrals, and drinking troughs on some slight topographic eminence above the usual high-water level within areas subject to seasonal inundation. The animals are brought here to be milked, to rest, to be treated for this or that, and to be loaded for sale. Underfoot one often sees sherds. There were Prehispanic settlement sites on these eminences, in some cases they are clearly constructed mounds. Hollows can sometimes be seen nearby, from which building materials were scooped. These hollows will have held very useful dry-season ponds, as they do now.

Ranching practice in Las Marismas, the estuarine marshlands surrounding the Guadalquivir River south of Seville, was clearly the model for early ranching in the tropical lowlands along the Gulf of Mexico. This system involved a movement between marshlands, grazed in the summer, and adjacent woodland, used in the winter. This required the use of horses—and fostered ferality. In the marshlands the cattle were left to themselves for long periods (Jordan 1989: 116). In tropical lowland New Spain the system became a movement between perpetually green swampy terrain accessible during the dry season and neighboring higher lands, part forested, part savanna, which green during the wet season. This system can be contrasted with that of the Spanish highlands to the north, especially Extramadura (Doolittle 1987; Butzer 1988; Jordan 1989). In the latter *ganado menor*, including swine, far outnumbered *ganado mayor*. Herding was done afoot and there was extensive transhumance but little ferality, because contact was maintained.

So when the médico calls the lush lowlands of Veracruz the Extramadura of New Spain, he is right only in a very general sense. It is important cattle country. However, it would have been more appropriate to apply that comparative designation to the upland ranching country of Tlaxcala and Cholula and to call the lowlands the Andalucia of New Spain.

One other condition facilitates animal husbandry in the Basin. The vicar notes that feed for mares, mules, donkeys, and horses can be brought from the swamps into town by blacks, all year. They will have had to get the green fodder from slightly higher areas during the wet season and from lower down in the dry season. The rhythm may be deduced, but the availability of pasture within a limited area throughout the year is clear.

Ranching in colonial Veracruz, and later, too, for that matter, is often called "extensive." There was some herding. How else would it have been possible to assemble the thousands of cattle that were chased each morning, when the miasmas were particularly bad, through the streets of Veracruz? Moreover meat was supplied regularly and abundantly for consumption in Veracruz. The *médico* was impressed by the *estancias* of cows and mares, implying some managed breeding. From the vicar's mention of cheese-making one may deduce considerable milking. But many animals did range freely within the Basin. The vicar tells about how some are brought down far from the ranch headquarters with *desjarretaderas*, poles with crescent-shaped knives at the end, used to sever hamstrings, and that the carcasses are left unused.

No *vaqueros* are explicitly mentioned in any of our early colonial materials. We may deduce black cowboys on the Ruíz property, but little more. It has been noted that the *vaqueros* of the lowlands were gaining a reputation for rowdiness by mid-century and were being denigrated by the better people (Jordan 1993: 92). One would think that the vicar would have noticed and decried such sinfulness.

The origin of *Jarocho*, the term that came to settle over lowlanders, especially Veracruzanos, seems to be lodged somewhere in all of this, perhaps in the history of the black population. It is not difficult to link the term to the *garrocha*, a pike used in Las Marismas and then in lowland Veracruz to manage cattle and modified with a knife to serve as a *desjarretadera* (Jordan 1993: 26, 77). The reputation for rowdiness certainly persisted—criticized by the better people and the uplanders, eventually turned back by Veracruzanos and adopted with pride as a distinction (Siemens 1990: 157–158).

Cimarrón

Cimarrón is a strong word, meaning "wild," "fierce," "savage," "untamed" (Santamaría 1978: 243; Williams 1955: 88, 396). It is useful to draw in the English equivalent, "feral", as well: "lapsed into a wild form from a domesticated condition" (Oxford English Dictionary 1982: 984). The word is most readily applied to animals, but in Latin America it has also been used to refer to escaped slaves.

Strictly speaking, feral animals should not be considered wild, "because they have been changed to some degree, at some time, by human intervention, and thus are not in an original state" (Lawrence 1982: 63). They have been tempered by:

> Two phases of selective pressure; one phase, domestication, can result in appreciable modification of anatomy, physiology and behavior and requires a high degree of dependence on man; and the second phase requires that the animal be capable of surviving in the wild independent of man. (Munton et al. 1984: 9)

It stands to reason that ferality would have its degrees; in the Caribbean of the early colonial period these were distinguished (Jordan 1993: 76–77). The wildest were the *bravos*, which remained in remote areas and were hunted, not herded. *Bravo* is the word the vicar uses.

In its physical form, the feral animal may become larger, through natural selection. This reverses the usual, and in the study of prehistoric animal remains diagnostic, reduction in size consequent on domestication (Clutton-Brock 1981: 22). Other physical characteristics are apparent—and attractive to anyone who wishes to hunt or capture and exploit such animals: "Low fatness, high conversion efficiency, good bone structure, resistance to disease, good foraging and mothering behavior" (Munton et al. 1984: 11).

On these and other, more impressionistic grounds, one can hardly avoid admiring *cimarrones* and *cimarronas*, unless one happens to be a cattleman missing animals or a slave-owner missing slaves. This positive image, an appreciation of freedom gained, is particularly strongly developed in North America with respect to the feral horse. It is a potent image out of the West, a symbol for it, especially *vis-a-vis* the city or the East (Lawrence 1982: 135). Fine stories are told about it: "The gentlest wagon horse (even though quite fagged with travel), once among a drove of mustangs, will often acquire in a few hours all the intractable wildness of his untamed companions" (Lawrence 1982: 136). Allusion to ferality has even, at various times, sold automobiles. Streaming herds of mustangs and rearing wild stallions are a staple of Western art and may be seen in television advertising from time to time.

One of the most engaging authors on ferality, Tom McKnight, observes that feral animals offer an offense, a challenge: Here is wealth running free that could be in one's own hands (1976: 62). Such an animal is certainly an important resource in Spain's colonies; one of the clichés of colonial history is that it was used with profligacy. Meat is abundant, one can afford to be discriminating and prefer veal.

Then, there is an important ecological aspect:

The establishment and permanence of ferality among cattle, as with other animals, is in large measure dependent upon the availability of suitable habitat. Most important in this regard is the difficulty of access for man. Country that is difficult for man is likely to be permissive for long-term ferality simply because man, the killer and capturer of animals, is excluded. Thus, feral cattle tend to have greater longevity where there is rough terrain or swampy ground or dense scrub. (McKnight 1976: 57)

The humid lowlands of the San Juan Basin offer such sanctuary. One imagines the animals moving about in it with care and light, elastic steps, emerging from the forested areas on the wetland margins and the "islands" within the wetlands to graze, at night, against the wind, watchful as wild turkeys, alert and swift as deer (Dobie 1980: 1–44).

The vicar may well have had a good deal of all this in mind: "Son bacas muchas y muchas cimarronas y muchos toros y muy brabos [There are many cows, many feral cows, and many bulls, some of which are fierce]" (A. Hernández 1571: 199; my translation).

Succession

Feral animals embody adaptive success in the early Central Veracruzan colonial landscape. They achieve a high degree of microenvironmetal articulation, which has always been critical for the successful exploitation of tropical lowland resources. The wetlands and surrounding forests become their sanctuaries. The herded animals realize the favor of the humid lowlands, particularly the wetland-forest *ecotone,* the critical zone *b.* We impute similarly effective articulation to the Prehispanic agriculturalists of the region. This, then, is the basic gist of succession: the Indian fades, and an impressive animal resolves in his place.

5. Maximizing Some Late-Eighteenth-Century Observations

"Excelentes pastos."
"Terreno . . . muy incomodo."

—Miguel del Corral 1777

In July of 1777 two Spaniards, Miguel del Corral ("coronel ingeniero en segundo") and Joaquín de Aranda ("capitán de fragata graduado y piloto mayor de derrotas de la Real Armada") completed a *Relación* on an extensive reconnaissance in Southern Veracruz, including deep in the midst of it a very brief sweep northward, an *Adición:*

> All the terrain inland from the coastal bar of Alvarado to some distance beyond the La Antigua River, with the exception of forested higher ground near the coast—in parts not more than a half a league in width—all the rest is gentle hills and savannas of excellent pasture up to the foot of the mountains. These grasslands are occupied by haciendas and ranchos of bovine cattle. The rivers are spaced well to provide the cattle with drinking water at intervals. They are lined with strips of slightly higher wooded ground that, together with the patches of forest in the savanna, offer shade for the cattle during the times of very high temperatures.
>
> From Veracruz and other landing places along the coast roads extend toward Mexico via Orizaba, Xalapa, and Iguatlán de los Reyes. These are good roads, except during the rains, when they become very bad.
>
> In the haciendas and ranchos of the area enclosed there were, in 1765 and 1766, more than 200,000 head of bovine cattle, more than half of which were feral, as well as 3000 mares and 5000 horses. Since these haciendas quickly deteriorate if not kept up and also are quickly reestablished, it cannot be said in what state they might be at the present, but judging from those seen between Veracruz and Alvarado, ranching is flourishing. (AGI, Mexico vol. 1864, f. 37–38; my translation)

Corral had already completed a lesser *Relación* on the lowlands in 1771. It provides a different view into the Basin:

> The coast from Veracruz to the La Antigua River is accessible and unforested; ships of any tonnage can anchor but there is no protection from winds out of any quarter. All of the terrain for twenty leagues into the interior is very uncomfortable for those who live and move about in it. In addition to the excessive heat, they must endure the plagues of mosquitoes, ticks, and other animals that are extremely bothersome and raise large sores on the legs unless one takes great care. (AGN, *Historia*, Vol. 359, Exp. 1,2, f. 12–13; my translation)

These brief observations, together with some fine maps and related materials (mostly attributable directly or indirectly to Corral), confirm the colonial succession, particularly with respect to ranching, as well as the outsider's view of lowland productivity. However, they also tap into a grimmer insider's view. Between the two views is a persistent tension, which we will come on again in subsequent chapters.

One hesitates to glide thus over two centuries, with only very general indications of change during that long time, but there is little recourse. Overviews of Mexican history tend to treat the colonial period *en bloc* (e.g., Meyer and Sherman 1979; Keen 1992). Undoubtedly there are trends and regional variations still to be traced; they must be left to other investigations.

The Documents

Archival research is almost enough to make one give up fieldwork: anticipation at the opening of a packet that a clerk has just delivered to one's desk, the feel of the paper, a key find, disappointments and surprising dividends—these are all very absorbing. The search represented here was begun in both the Archivo General de Indias (AGI) and the Archivo General de la Nación (AGN) with guides to the maps in the holdings—a geographer's approach (Torres Lanzas 1900; AGN 1979). The resulting complement of reports, letters, procedural documents and maps is quite coherent but not very substantial for purposes of this investigation.

Late in the eighteenth century the king of Spain, Charles III, and his viceroy in New Spain, Antonio María Bucareli, as well as the *visitador* José de Gálvez, were much concerned over the defense of

New Spain against the English. Keeping troops in the lowlands was difficult; death and desertion took a heavy continuing toll; funding and supply were constant problems. At various times there were special commissions on questions of defense and fortification. Better information on roads and the coastline was ordered; Corral's 1771 *Relación* was one of the responses. Improvements in fortifications were attempted; a foundry for bronze artillery was planned for Orizaba, but the project came to nothing. It was determined that the feasibility of a shipyard on one of the rivers of the Gulf Lowlands should be examined. This was the charge given Corral, which resulted in the 1777 *Relación* and its maps, but it also came to nothing (Siemens and Brinckmann 1976). All this did at least leave a residue of documents, now grist for our mill.

Corral was born in Aragon in 1731, took up a military career specializing in engineering, and served with distinction in Spain under Bucareli, when the latter was a field marshal. He came to New Spain in 1763 and was soon involved in various commissions, including surveys of the port of Veracruz and its hinterland. The information obtained during that research is basic to the materials considered here. Along the way Corral was given various important responsibilities, including eventually the governorship of the intendancy of Veracruz. He became a famous Veracruzano (Pasquel 1963: xii–xvii).

The first of Corral's contributions to the geography of Central Veracruz that is of interest here is the 1771 *Relación* already referred to. It is coastal reconnaissance with a shipboard purvue, complemented by a simple tracing of the Mexican Gulf Coast from Alvarado northwestward almost up to the mouth of the Río Nautla (AGN, Historia, Vol. 359, Exp. 1,2, f. 1–17). Both are focused on questions of maritime access and defense and have an ominous feel about them.

Reports to the home office were routinely introduced by protestations of diligence and devotion; Corral's preamble to the 1777 *Relación* is a beautiful example. However, the written documents and the maps do actually show extensive fieldwork and geographical competence. The author focused clearly on the objectives he was given and reasoned plausibly, providing a consequential guide to decision-making. Both Bucareli and Gálvez commended him highly. He and his various collaborators "write" the Basin landscape and its surroundings for us in the late eighteenth century.

The 1777 *Relación*, it should be noted, is not part of the series of *Relaciones Geográficas* elicited in Mexico and Central America between 1740 and 1792 by the Bourbon kings (AGI, Indiferente de

Nueva España, Legajos 107 and 108, as cited in West 1972). I had for years been tantalized by West's notation that there was a substantial 1744 *Relación* in this series for Veracruz Nueva, i.e., the town that grew into the port one sees today. When it was finally possible in 1996 to look at this *Relación*, it turned out to be mostly a lengthy and exhaustive census of the town, plus an inset that was almost a taunt: a description of a mapping expedition into the surroundings. The map has not been found.

The 1777 *Relación* is organized into five chapters: (1) a geographical description of what is now Southern Veracruz state, with additional information on the isthmus, and the sweep northward, the *Adición*; (2) a survey of the region's forests and lumbering activities, with estimated prices of wood included; (3) a reflection on the advantages and disadvantages of two possible sites for the shipyard, with a recommendation on which one should be chosen: Tlacotalpan; (4) proposals on the fortifications that would be needed to secure the investment; and (5) a survey of the military forces and installations available in the area for its own defense.

The *Relación* has a curious sequel. In 1781 Bucareli's successor, Martín de Mayorga, needed to inform himself about eastern defenses, since Spain was then at war with England. He asked for information out of the archives but none could be found; the *Relación* was lost or hidden. His bad relationships with the Spanish court and the governor of Veracruz, as well as a junta convened on the question of the fortification of Veracruz (which included Corral), may have had something to do with this. In any case he commissioned another survey, now in the AGI, but it is less elaborate and incisive.

The 1777 *Relación* was published by the Editorial Citlaltepetl of Mexico in a volume called *La Costa de Sotavento*—part of a *Colección Suma Veracruzana*, no less. Better, perhaps, not to be published at all than to be published like this. The volumes of the series are sometimes more and sometimes less relevantly introduced by the Editorial's *director-gerente*, Lic. Leonardo Pasquel. The paper is poor, the bindings weak, the illustrations execrable, and there is a blithe disregard for citation throughout. However, the volumes are often based on what is obviously good material and hence it must be admitted that they can be quite useful.

The *Relación*, and the *Adición* within it, have their *pinturas*. The first is paralleled by a map of Southern Veracruz, which covers the northern limits of the Papaloapan River basin southeastward to the Isthmus of Tehuantepec (AGI, Planos de Mexico, no. 329). It is a gratifying piece of cartography, richly colored in greens and browns, and quite impossible to reproduce properly here. A fulsome legend provides good terms of reference but also quickly induces regret: this map, it says, is a reduction of a larger map on which there were important details. That larger map has not been found.

Corral and those who worked with him eventually also produced two versions of a map of Central Veracruz, paralleling the *Adición*. The more detailed and finely drawn of the two is presented here (figure 5.1). It has a long legend from which one may deduce that the main point was a survey of the roads of the region and that the fieldwork on which it is based is that same series of field seasons in the late 1760s that produced the 1777 *Relación* and the cartographic masterpiece.

Intimations of Humboldt

Considering the Corral maps, appreciating their fabric, the fine lines, the routes, and the toponymy puts one in mind of the savant's map of Central Mexico (figure 5.2) and the epochal *Political Essay on the Kingdom of New Spain* (Humboldt 1911). There is a relationship.

Humboldt claims exhaustive cartographic research on the lowlands, including consultation of Corral's "itineraries"—presumably field materials. He could hardly have seen the maps we have referred to here. If he had, it is unlikely that his essay on New Spain would have shown the serious lacunae and even misstatements that it did on the intendancy of Veracruz (Siemens and Brinckmann 1976: 270). He claims repeatedly to have made the latest in topographic measurement himself and, with the aid of his draftsman in Europe, he presents his results very convincingly. The landmass is positioned correctly, orientation is as it should be, and the margin is calibrated to latitude and longitude, lending additional authority! This was the best geographical work on Mexico up to that time (cf. Thrower 1990: 30); it would intimidate others for many years.

In his map, in his remarkable topographic profile of Central Mexico, and in his essay, Humboldt posits various important, large ideas about the slope from the summit of the Sierras to the sea, notably his influential altitudinal scheme. He is credited with having rendered the third dimension better than anyone before him (Beck 1985: 242). He also gives us a breathtaking assessment of the lowlanders, which is useful as a target in any consideration of prominent foreign views of the people of the Mexican tropics (Siemens

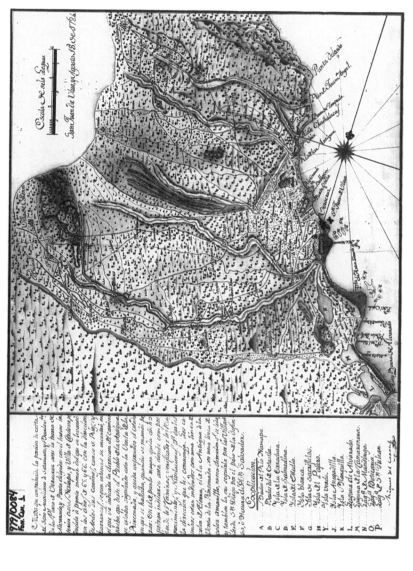

Fig. 5.1. A map of Central Veracruz in 1784 by Miguel del Corral (AGN, Fomento Caminos, Vol. 1, f. 91).

Fig. 5.2. Central Veracruz by Alexander von Humboldt (excerpt from Humboldt 1969: 9).

1990: 28–29, 192, 148–149). But we cannot resort to him for a reading of the Basin; our analysis is on a different scale.

The "Political Unconscious" in the Cartography of Central Veracruz on the Eve of Independence

Accuracy in cartography has grown since the maps of the late sixteenth century. Objectivity is aspired to more and more urgently. Corral argues deliberately, point by point, as though science were being marshaled in aid of administration. He letters features and refers to them systematically. His north is not yet up, but the heights and coastal shapes are close to what we know they should be. Latitude and longitude are cited, although not as elegantly or punctiliously as in cartography by Humboldt, who made a very deliberate point of having measured accurately. This had become a "talisman of authority" (Harley 1989: 10). Burgeoning science, in turn, makes the late colonial reality more accessible to the twentieth-century observer, helping us, in our way, to take possession of the lowlands.

Ethnocentricity has virtually erased Indian toponymy; this is the great silence of these maps. Power is perhaps no longer as crassly expressed as in the legal maps of the late sixteenth century, but then colonization is a done thing. The Spanish design is embedded in the hierarchy of symbols, particularly those used for settlement on the 1777 map. The *cabezera* (*cabecera*) or *villa* is in the first place. This would be a predominantly Spanish town, the regional center of administration. The *pueblo* is second, a smaller town, still largely Spanish. Rural settlement is dominated by the third entity, the hacienda. This term has replaced *estancia*, which is common in the earlier colonial literature. The *ranchería* completes the series—a small rural settlement. Humboldt is fairly strictly hierarchical as well. One may infer the presence of some Indians in the pueblos and rancherías, but the scheme is Spanish and only an occasional place-name echoes what was before.

Considering the importance of secrecy regarding possessions and strategy in colonial administration (Harley 1989: 59–65), we must, after all these years, be seeing in the Corral maps "declassified" documents. They were withheld from foreigners, probably including Humboldt—all his excellent contacts notwithstanding. Spain's hold on the lowlands and indeed New Spain is slipping. Information on the realm is being updated and guarded, and administration is being reformed, but this is, if not too little, certainly too late.

The More Prosaic Content

On the Corral maps, particularly the 1777 map of Southern Veracruz, the third dimension is dramatic: serried mountains around the lowlands with the volcanoes emphasized, a piedmont gashed with *barrancas*. However, both the 1777 and 1784 (figure 5.1) maps are still very impressionistic topographically; in fact, in this respect they remind one of various sixteenth-century *pinturas*. On Humboldt's map there is not a great deal more systematic topographical information, but there is altitudinal shading, which shows a con-tour line-like ascendance westward toward the volcanoes. The canyons are less comprehensive than on Corral's map, indicating no great field familiarity with this region, although Humboldt did pass through it on his way home. His great ponderous profile of East Central Mexico (Figure 5.3) is not very enlightening on the topography of Central Veracruz; it is more or less one steep convex slope, the vertical scale is such that a stone dislodged somewhere east of the top of the escarpment would thunder right down to the lowlands, which are themselves reduced to a slight thickening of the base line. However, its heights have been measured and in the margin Humboldt outlines his altitudinal scheme.

We see Humboldt's map and profile as published engravings; Corral's maps come to us in their original drawn and painted form. This puts the former at an advantage; they look more authoritative. The nature and effects of engraving will be referred to again.

Hydrographically, Corral's maps give us little more on the Basin than a crude sketch of the verbal point already made in the late-sixteenth-century documents: a number of streams join the main one, a substantial multichanneled affair, which then flows in a straight channel northward into the La Antigua River. Humboldt, more elegantly, does the same thing. *Laguna Catarina* takes many forms in the materials that underlie our analysis of the Basin. There is no lake at all on any of the relevant Corral maps. Humboldt shows only a fragmented minimum. Once one has seen the full extent of flooding in the Basin during a year with normal or better rains, one does not easily forget the open water. Corral and Humboldt's maps are dry-season productions; by the time the lake has taken its full extent, all sensible investigators of tropical lowlands have retreated upslope.

The Corral and the Humboldt maps are sifted over with vegetational motifs: clumps of bush and grassy knolls on the latter, and most curiously, a tree and its shadow on the former. One scans this tantalizing cover to see if there is not some symbolization, particu-

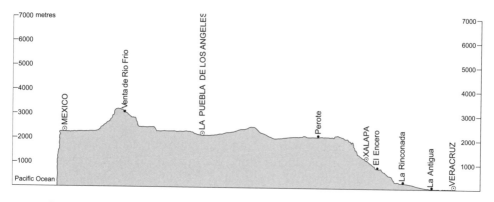

Fig. 5.3. A measured profile from Mexico City to Veracruz (adapted from Humboldt 1969: 12).

larly some systematic differentiation between savanna and forest. There is none; this is decoration, like wallpaper.

Corral's *Relación* sweeps from Central into Southern Veracruz; the descriptions in it of the various forests are its best material. Tropical hardwoods for ships' hulls are obtained from Medellín southward, on the margin of the world described in the sixteenth-century documents from Old Veracruz, and conifers for masts and spars come from the higher areas to the west and southwest.

At one point Corral ruminates about cutting practices and criticizes the loggers quite sarcastically for knocking down unwanted or immature trees and never thinking about replanting. He even suggests a way of enforcing care in cropping. For a moment he has become an agreeable contemporary.

Agriculture is hardly apparent on any of the maps. There is some mention of it in Corral's description of the Papaloapan Basin, from which the town of Veracruz is supplied. The growing of crops is still an Indian thing. Both agriculture and Indians are even more diaphanous in these materials than they were in the descriptions by the vicar and the médico.

Point, Counterpoint

In Corral's 1771 *Relación* the vulnerability of the Veracruzan coast is the main concern. He provides a shipboard view of what lies behind the beach just north of Veracruz. In a few sentences he sketches out the menace of the Basin. It is difficult to move into the interior. Those who live there or have to pass through it often, as someone from the region must have informed him, find it most uncomfortable. The vegetation is dense; it's very hot and one is plagued by insects.

The implication is that this menace will also deter invaders. American soldiers were able to land easily enough on the beach just south of Veracruz in 1847, but they found out how difficult it was to move into the interior toward Jalapa, and wrote about it eloquently (Siemens 1990: 133–147). The basic problem in these parts until the new road was built at the end of the eighteenth century, and even then, was how to get safely from the coast to the plateau. It was not so difficult along the main roads in the dry season, but very difficult after the onset of the rains. The coastal lowlands *per se* were the colony's best defense.

Corral also gives us a very different view of the lowlands, the view represented in the *Adición* to the 1777 *Relación*. It is sweeping, interested, and peninsular, rather like pronouncements on productivity in the accounts of the early Spaniards coming to these coasts. For some reason it has become important for the author to take an official view, to note the continuing desirability of these lowland regions of the colony and the necessity to defend them. There are gentle hills and savannas, excellent pastures, plenty of drinking water and good shade. Ranching, the great colonial introduction on this coast, is flourishing. It would diminish dramatically in the coming years of insecurity, at least as far as numbers of cattle are concerned (González Jácome 1997: 107).

Cimarrones Again

Corral estimates that there are 200,000 bovine cattle, plus 3000 mares and 5000 horses, in the lowlands north of the Papaloapan Basin. It is not clear just how one should take the distinction between mares and horses. Corral did not mention mules in this connection, but he carefully enumerates them on haciendas to the south. And he has surely taken the donkeys completely for granted. His estimate of bovine cattle is roughly in line with the estimate of 150,000 given by the médico in 1580, since the area he enclosed was probably somewhat larger.

Corral gives us pause with his thought about the potential impermanence of a ranching operation. It can deteriorate quickly if not kept up and can be reestablished just as quickly. One would infer that pastures in any given locality can be opened up relatively quickly by clearing and lost to woody growth again quite quickly if not regularly "cleansed," as the present saying goes. This caution and the comparable numbers at the two extremes of the colonial period indicate that over these centuries ranching in general within this region may have been in a kind of dynamic equilibrium. The short-term changeability needs to be kept in mind when considering the actual use of any given land grants, as well as broad issues like deforestation and reforestation in the tropical lowlands when observed at any given time.

More than half of the 200,000 bovine cattle in Central Veracruz are feral. From a comment in the 1771 *Relación* it is apparent that in the Papaloapan Basin to the south an even greater proportion of the cattle are *cimarrones*. These are hunted once a year, to supply meat markets in Veracruz, Cordoba, Orizaba, and other towns. Ranching remains rudimentary, or "extensive" (e.g., González Jácome 1997: 111), but perhaps less so than it was in the early colonial period. Fresh, i.e., sun-dried, meat and not just hides are taken from the feral cattle that are hunted down.

They tantalize us again, the impressive beasts coming out of the forest at night. They remain the clearest indicators of the nature of colonial land use in this region, the ecology of that land use, and the stability of the system. The animals themselves best realize the favor of these lowlands.

6. Appreciating a Naturalist's Rendition of Central Veracruz in the Nineteenth Century

"Dunkle Waldungen, allmählig ansteigend."
(Dark, gradually ascending woodlands)

—Sartorius 1859: 1

For the long parade of non-Spanish foreign observers that enter newly independent Mexico, Ortega y Medina, the eminent Mexican historian, has mordant remarks: they are tireless globetrotters, tramps, a motley, garrulous caravan of adventurers in search of easy gain, new sensations, and romantic spasms, which will liberate them from the fatigue, the boredom, and the limited opportunities of the old world (1955: 12, 20, 26).

The parade includes diplomats who have come to establish new relationships; their observations are often condescending. Representatives of investment and trading companies are hurrying in; they sense new opportunities, which would prove difficult to realize. Merchants have come to profit, hugely, and they mostly keep ledgers rather than journals. Other travelers note the sallow complexions and dissolute ways of these servants of Mammon. Writers and artists produce some arresting imagery. American soldiers make their observations, too, as they fight their way inland in 1847. They are repeatedly ordered across country; they describe the terrible heat and the disconcerting vegetation on the dunes and the forested hill land. Fortunately, there are also a number of observers who have the instincts of naturalists.

Carl Sartorius is one of those; his altitudinally articulated lowland landscapes are a satisfying sequel to the terrain in the colonial documents. They are in an idiom that is close to our own, and they help us to clarify the earlier materials, indicate changes in settlement and the environment in the interim, and provide a basis from which to consider the substantial changes that would come in the next century.

The common early-nineteenth-century route of entry and exit is the impressive road constructed during the last years of the colonial period from Perote down through Jalapa and Rinconada, rounding the Basin and going on as far as Santa Fe, to be followed finally by a rather ignominious struggle through the dunes and then a pleasant stretch along the beach into Veracruz (Siemens 1990).

Coming or going the travelers are in a great hurry to get through the lowlands proper, past the Basin. From the swamplands emanate the evil airs that bring disease. They are considered particularly dangerous during the rainy season, so one arranges to arrive in the dry season if one possibly can, but hurries through them regardless, just in case. Up at the altitude of the first oaks the danger of infection with yellow fever abates. We thus get the terrain below that described mostly in passing, but fortunately not all the observers were in a hurry.

Two main predispositions are apparent: First, the nineteenth-century observers are impressed with the luxuriance of tropical nature—not without a certain appreciation of its malevolence in the lowest of the lowlands. That green band of lush vegetation a little higher up, which they can see from the decks of their ships, out there beyond the dunes and below the peak of Mount Orizaba, will repay them for the difficult crossing. These are the "dark, gradually ascending woodlands" that Sartorius sees (1859: 1).

The other main bias relates to the lowlanders. They have proved quite inadequate to the challenge of their environment; in fact, they have been victimized by it. Humboldt had already described the luxuriance and dealt at length with the dangers of the lowlands; he puts this idea squarely as well: the lowlanders could be very productive if "their laziness, the effect of the bounty of nature, and the facility of providing without effort for the most urgent wants of life, did not impede the progress of industry" (Humboldt 1911: 253–254).

This is a hardy, persistent dyad: luxuriant nature, inadequate man. It arises in Classical times, is restated memorably by Humboldt, and is adopted more or less directly by most of the observers who come this way in the nineteenth century. It is still detectable in some of the basic literature on development of tropical lowlands in the second half of the twentieth century (Siemens 1989b).

The "Serious" Observers

As it happens, most of the naturalists that follow Humboldt into Mexico are German. Ortega y Medina was particularly interested in, and critical of, Anglo-Saxon observers and perhaps should have restricted his judgments to them, but he does offer a comment on German observers too, excepting them from his sharper judgments: "German travel literature on Mexico is highly valuable for its sincerity, its stubborn objectivity and the benevolence of its judgments" (Ortega y Medina 1955: 126; my translation).

One *does* resort with alacrity to the observations of the earnest German traveler of the nineteenth century, no doubt about it. They seem so reliable. One is tempted to say that it is just the density of the prose, but there is more to it. There are the consequential developments in geography in German universities of the time, the elaboration of regional and systematic studies, as well as the very considerable reputation of Humboldt, to all of which nineteenth-century German observers with scientific inclinations were heirs.

Ortega y Medina thought that the German travelers actually suffered under the example of their illustrious precursor. He was obligatory spiritual fodder—not only for the Germans but for anyone else coming from Europe. His was still the measure of things written about Mexico, and that measure will often have been intimidating. Humboldt had been there for about a year, from 1803 to 1804; had searched out information on the viceroyalty; had drawn, mapped, and described what he saw or was made aware of; and had formed opinions that were extremely influential in Mexico and abroad. The truly well connected traveler in the early decades of the nineteenth century came with a letter of reference from this last of the renaissance men. Many felt constrained at least to emulate his comprehensiveness and incision, while others thought they might update him or at least complement his monumental work. Some disavowed any such intentions in their prefaces but then went on to do rather well nevertheless.

As for objectivity and benevolent judgments, one has to wonder how much of the German literature Ortega y Medina actually read. Most of the authors had appropriated Humboldt's prejudices and given them a few turns of their own. They can be quite as arrogant as he, and their prose is as sprinkled as his with lazy lowlanders, Indians of limited intelligence, and barbaric rituals. Carl Sartorius shares the predispositions of his compatriots to one degree or another but has various further personal inclinations that serve us rather well. He easily takes his place with the vicar, the médico and the naval engineer, Corral, as an engaging interpreter of Central Veracruz.

Observations at Length

Sartorius organizes some of his material along itineraries; they seem fairly clearly to be the syntheses of many journeys. He also tells us that he habitually went on horseback in order to be able to digress easily. He speaks not just as an adventurer, but as a resident. He comes to Mexico in 1824 at twenty-seven years of age and stays, except for a three-year return to Europe, until his death in 1872. At the outset he successfully manages various mines, but he then acquires land from the huge hacienda Acazónica, in the attractive midslope region south of Jalapa, near Huatusco. He develops it into a very respectable and innovative holding, called El Mirador.

Out of his forty-five years of residence at this place, and the not inconsiderable travel in other regions from time to time, comes his main work, a portrait of a country. An adopted country? Hardly. Sartorius nowhere indicates that he considers himself to have become a Mexican; he remains fairly clearly a German abroad and evidently hoped to retire at home (Sharrer 1980: 41). He presents Mexico as a potential homeland for others of German background but expects that they will remain German at heart, too.

Before Sartorius's book can be appreciated, there is a bibliographical thicket to penetrate. It was first published in outline within a slim treatise on German emigration in 1850 (Sartorius 1850), and then in its full German version some years later (Sartorius 1859). An English version was evidently available in 1858; it was reprinted in Stuttgart in 1961 (Sartorius 1961). It is quite careful and competent on the whole, but rather wooden. The translation was done by an anonymous someone whose mother tongue was not English. The book was then edited by a Dr. Gaspey, who has several things to answer for. He excised significant passages, following criteria that are not entirely clear.

Two Spanish versions have since appeared, one translated from the early English version (Sartorius 1991), the other from an English reprint of 1961 (Sartorius 1990). Both have valuable introductory notes but show the same gaps as their English antecedents, and of course are twice removed from the original.

A copy of the English version of 1961 was this author's introduction to Sartorius. It has no index, and so in the course of various protracted analyses its edges frayed and its binding parted. Subsequent readings have been tempered by repeated considerations of the German original, which fills in Dr. Gaspey's excisions, and by Sartorius's other publications. The most substantial addition to the work at mid-century is his essay on the state of agriculture in the Huatusco region, written in 1865 and published in 1870, two years before his death (Sartorius 1870). This seems a kind of testament, still positive but tempered greatly by the intervening years. Although it treats of terrain well upslope and to the southwest of the Basin, it is very relevant in certain of its broader insights. All of this is put into context by Pferdekampf's well-considered survey of the German presence in independent Mexico (1958), Sharrer's thesis on Sartorius and his hacienda (1980), von Mentz's analysis of the nineteenth-century German view of Mexico (1982), and her introduction to the recent Spanish publication of Sartorius's main work (1990).

Predispositions

During his student years Sartorius formed strong libertarian ideals, for which he became a fugitive. Those who professed liberalism in Europe and in Mexico espoused freedom, elevated the individual, and laid great store by private property. They identified with the elite but decried feudalism and, indeed, communal or corporate administration of property as inimical to progress. They showed little sympathy for indigenous people and resorted to a pronounced environmental determinism. Traces of all of this are easy to find in Sartorius's work.

It has been pointed out that like other observers of a liberal persuasion, Sartorius had only a partial understanding of the social reality of his time (von Mentz 1990: 13–27). This assessment will need to be qualified.

Sartorius and his friends had dreamed of establishing a German utopian community abroad when they were still quite young. Such a community was indeed attempted near Sartorius's hacienda in the 1830s but soon had to be given up. In spite of this serious disappointment Sartorius remains convinced that German immigration into Mexico is feasible and desirable. Near the end of his life he affirms it still, but with qualifications: he advocates that conditions should be created such that families would be stimulated to come and take up private properties—a tall order, influenced by what Sartorius had found out about agricultural settlement in the United States (Sartorius 1870: 195–196).

Advocacy of colonization is coherent with his views of tropical nature and tropical people. A fine array of minimally used resources awaits hard-pressed European folk. The Mexican population will never be able to effect the development of its own country. It needs immigrants with knowledge, industry, and high morals. Mexico can be a place in which the German spirit can triumph.

Sartorius elaborates on what it is that will make Germans the salt of the Mexican earth (1850: 55). They are cosmopolitan, able to integrate easily into foreign societies. They have a profound humanity (whatever that may mean), which allows them to find and rec-

ognize the true and the good in foreign cultures. Mexicans are volatile, Germans calm; the two harmonize and intrigue each other—
an echo of an old germanic inclination southward? The Mexican is
not able to overcome his frivolity; gambling and women dissipate
his capital. The German, through right living and good housekeep-
ing, maintains his holdings and multiplies them. This, Sartorius
points out, is well known in Mexico. Critics would eventually take
umbrage at this Eurocentrism and arrogance, as might be expected
(e.g., von Mentz 1990: 12). The prospective German immigrant, he
suggests, should make his move with confidence, assured of a good
reception.

Such imagery is widely held to this day, as anyone with a German name living or traveling in Latin America knows. It is imputed, whether deserved or not, as a matter of course.

Sartorius studied law and philology but obviously has a wider
curiosity. In Mexico he behaves as an amateur in the best sense of
the term; he is indeed a naturalist. Alongside all the other urgent
and useful things he does, he explores, observes, collects, and draws
from nature too, it seems, but not much of this artwork is published. He contributes in various ways to the proceedings of scientific societies in Mexico and the United States. His principal interest is botany, but he is important to us also for a range of other
perceptive natural environmental observations, as well as a number of rather good verbal sketches of material culture and liveli-
hood systems, and for his fascination with ancient remains.

The direction of Sartorius's view on the lowlands proper, the
lands that interest us, follows from the location and the name of his
hacienda, El Mirador. The principal residence, still there today, is
set on an eminence within the most attractive of the bands that line
tropical mountainsides: the temperate zone which shares plants
from above and below, is not too unbearably hot nor too cold, and
offers safety from infection by various maladies. This is his favored
place. Its surroundings are the region for which he advocates German colonization.

The name itself is interesting; it means a vantage point. Traditionally, this could be a kiosk on the flat roof of an urban residence, a balcony or gallery overlooking a street or a working area,
open arches let into a church tower all around, or a viewing platform rising up over the roof of the main house of a hacienda; it is
always a place from which to survey the scene. One can think of it
as a *panopticon*, a place from which to extend authority, particularly if it surmounts the master's house (Siemens 1994). But it is
just as well a place of recreation, from which to enjoy the landscape

and the sky, from which to take strength and to plan. This is fairly clearly the sense in which Sartorius applies the term.

Although Sartorius's view is from a midslope vantage point, and he does not propose colonization in the lowlands proper, he has obviously not been afraid to go downslope. He has explored the lowlands, no doubt to satisfy his own considerable curiosity, perhaps in aid of his own early search for land, but surely also in order to be able to present the Central Veracruzan region fully to prospective colonists, to distinguish the various regions as to their promise. This involves saying something about yellow fever, the great horror of this coast. He does this succinctly, relegating the disease to the lowest areas, which one can pass through quickly, and he restricts it to the summer months. Moreover, one can avoid yellow fever even in the summer if one diets carefully and stays in the shade. Some other foreign observers of the time have similar advice (Siemens 1990: 82–94). In any case, Sartorius does not hurry through the lowlands, eyes averted. He deliberately describes activities, inhabitants, and environments, and what is very important for our analysis, he looks at the relationships between adjacent environments.

Sartorius enjoys midslope country life; he is disgusted with European and Mexican cities and praises the sky and the forest around El Mirador. He admires the *ranchero* and the *hacendado* who lives on his land. By reason of background and actual practice, Sartorius is neither ranchero nor hacendado, but he is a man of the land, almost hedonistic in that regard, and as he grows older he becomes a rural savant. Learned and famous visitors, including the Emperor Maximilian, come to his home in order to share its ambiance and to consult its owner on many issues, sometimes to join in an evening of song.

The various characteristics and inclinations of this author facilitate the pursuit of the theme of favor and allow us to trace landscape succession through the nineteenth century; no other observer of the time serves nearly as well. His aura of trustworthiness does no harm at all. His racial and social judgments are often hard to take, but his observations of material culture are both deft and ample. His scientific inclinations finally provide us with some ecology. His intrepid travel gets us into places where no one else will take us. His penchant for the rural life is useful when so many authors, both native and foreign, rush back to the cities. Sartorius's conviction regarding colonization does make him evasive in some respects; he stresses the positive, but that does help to engage the reader. His major book and all his other writing, too, is tilted from the Sierra eastward, which is exactly what we need.

The Magnum Opus

Sartorius sets an apologetic preface: the book is not a travel account; it cannot provide the diversion of such literature. But neither is it systematic or scientific; it is really only a series of sketches. Mind you, he has lived in this magnificent country for many years and does know, from personal observation and experience, that most of what has been written about it by others is wrong. Anyone who wants the best, says Sartorius, making the obligatory connection, must resort to the classic: Humboldt's essay on New Spain. Anything that he, Sartorius, can offer must be seen as "ornamental carving and fluting of the great master's strictly correct edifice" (Sartorius 1961: vii). For our purposes Sartorius himself is structural and the other observers of his time are decoration.

The title of the first English edition of his main work, *Mexico, Landscapes and Popular Sketches,* describes the content well. The book first treats Central Mexico's main altitudinal zones, from both coasts to the plateau, emphasizing the eastern slopes. These chapters are very satisfying indeed. Then we have the inhabitants presented to us, by the classes commonly recognized at the time. This is much less palatable. It fits what has been said about the contrast between the impressive way in which landscape and the natural world is treated in the work of nineteenth-century naturalists, over against their very much less impressive ethnography (Eisely 1961). Fortunately, Sartorius comes around toward the end of the book to economic activities, including an altitudinally articulated discussion of agriculture and ranching. We will appropriate the two endpieces rather energetically but step lightly over the middle.

Sartorius approaches Veracruz from the sea and then deals in turn with the savannas, the evergreen forests, the highland pines, and the plateau, including the analogous drop toward the Pacific. This organizing idea has its antecedent in Humboldt's altitudinal scheme. He had contemplated it first in the Andes, before he arrived in Mexico. In Central Veracruz he saw it again:

The admirable order with which different tribes of vegetables rise above one another by strata, as it were, is nowhere more perceptible than in ascending from the port of Vera Cruz to the table land of Perote. We see there the physiognomy of the country, the aspect of the sky, the form of plants, the figures of animals, the manners of the inhabitants, and the kind of cultivation followed by them, assume a different appearance at every step of our progress. (Humboldt 1911:251).

Various travelers appropriated the scheme uncritically in isolated statements and exaggerated it:

This changing vegetation is a barometer which, in Mexico marks the ascent and descent as regularly as the most nicely-adjusted artificial instrument. So accurately are the strata of the vegetation adjusted to the stratas of the atmosphere which they inhabit, as to lead the traveller to imagine that a gardeners hand had laid out the different fields which here rise one above another upon the side of the mountain that constitutes the eastern enclosure of the tableland . . .[a] specimen chart, where all the climates and productions of the world are embraced within the scope of a single glance. (Wilson 1855: 80–81).

It is not as clear as all that. Any reasonably observant traveler will notice some environmental gradation along a tropical mountainside, but it takes considerable effort, with maps and botanical guides in hand, to arrive at a comprehensive understanding, what with highly accidented topography and the vagaries of settlement and agriculture. The scheme is an abstraction, but a very useful one.

There is a well-known rendering of Humboldt and his companion, Bonpland, in a building somewhere along the Orinoco, surrounded by plant specimens, dead animals, and field gear—the two of them properly dressed in coats, vests, and ties! Among the litter is a pair of mercury barometers (e.g., Pratt 1992: 118). They were used to measure heights, which had become a great imperative. Travelers argued over them and corrected each other. Humboldt measured altitudes from the Central Plateau down to the Gulf, and the profile that resulted has already been referred to (figure 5.3).

In 1959 German geographer Carl Troll published his comparative study of tropical mountains (Troll 1959). He tells us it is meant to commemorate the centenary of Humboldt's death and links it explicitly to the altitudinal scheme first elaborated by Humboldt in the Andes. Troll's prose and profiles remain one of the most stimulating approaches to the Andean physical environment. He generalizes the surface, reduces the vertical scale just enough to keep his diagram mountainous but to prevent precipitousness, and then adds very apt symbols for the plant communities. Sartorius does not provide us with anything quite like this, but his prose suggests such profiles.

It has always been difficult to do justice to the topographic subtleties of lowlands. The profiles that West and his colleagues generated for their study of Tabasco, and many other aspects of that study, too, have long been an inspiration (West et al. 1969). Again, the surfaces are generalized enough to be diagnostic but not so much as to become implausible. A fringe of vegetation is applied to the topography with a set of symbols—left realistic enough to be immediately suggestive. This, too, is in the background of our visualization of Sartorius's landscapes.

Sartorius's own "sketches" and "pictures" are literary; there is not even a map. It seems that after Corral and Humboldt, no one needed or dared to try again for a long time. The book is graced by eighteen engravings of scenes from the work of the artist Moritz Rugendas, which raise an intriguing antiphony within the text.

The artist regarded his own work as inspired by a love and admiration for science; he did in fact illustrate the works of various European scientists and was commended by Humboldt himself for his ability to capture the physiognomic aspects of landscape (de la Torre 1985: 42, 76). He visited Mexico from 1831 to 1833, a year of which he spent in the Huatusco region, as a guest of Sartorius. At some point he visited the San Juan Basin, as well. In 1850 Sartorius wrote him to renew a friendship after many years, to remind him of pleasant hours at the hacienda, of excursions into ravines and mountains, and to tell him about his plans to publish a book on Mexico, adding "You would be the person who could best illustrate it" (de la Torre 1985: 77–78).

Rugendas sketched first and then painted later; he did not like to paint (de la Torre 1985: 38), but that seemed to be necessary in order to formalize and to sell. The drawings are consistently the most engaging; his lines are strong and his perspectives sure. These images would have made the best illustrations! The paintings are broad-brush, sometimes done repeatedly in different moods, and often quite perfunctory. On a few of them Rugendas resorts to finer brushes and takes care with vegetation—such paintings one could consider illustrations.

The images, drawn or painted, had to be engraved before they could be published. The uncompromising cuts into wood or metal add sharpness and authority: the vague is made definite. Textures become almost palpable; forms can be presented with precision or at least fine lines. And shading can be used as one might play a spotlight on a stage. But, in the transformation, material is assembled,

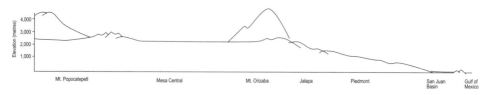

Fig. 6.1. A schematic profile of Central Mexico.

rearranged, made more comprehensive and, indeed, didactic. Such images have become valued antiques. Dismembered or whole editions of one or another of the nineteenth-century artists can in fact be found on any given Saturday in the antique market in the Zona Rosa of Mexico City.

Some of Rugendas's titles and drawn details show awareness of altitudinal zonation in vegetation—largely the results, no doubt, of interaction with Sartorius. However, he does not really seem to grasp or care about the abstract scheme *per se.* He decorates it but does not illustrate it; he is composing pictures. He is very much more in tune with the way that Sartorius and other authors of the time, especially Fanny Calderón de la Barca, verbally romanticize and thrill to the picturesque. And when his images are engraved, romanticism is positivized.

Sartorius read against a background of Troll and West, and supplemented here and there by the observations of others in the "garrulous caravan," evokes profiles at various scales, as did the vicar and the médico. These are not measured but schematic profiles over which one can extend a historical process, place ecologies, and undertake heuristic flourishes of other kinds (figure 6.1).

The Coastal Lowlands

The coming ashore, for Sartorius and for many others, is disagreeable. The landscape looks desertic; one uses arabic metaphors (figure 6.2). Not only that, the town among the dunes is dangerous; many contract yellow fever there and die. But a band of greenery behind it promises the luxuriance that will compensate the traveler for the voyage. The skyline is graced by an "immaculate cone" (Taylor 1850, 2: 324), or other words to that effect:

> Regarded as a whole, the coast has the same features, whether we trace it to the south or north. Everywhere we find a narrow level tract of coast, not many miles in width, then a gradual

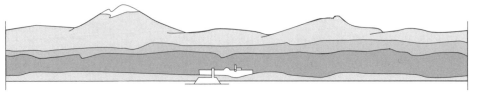

Fig. 6.2. The banded mainland of Central Veracruz as seen from the sea.

ascent by gently inclining slopes to the spurs of the mountains, and finally to the highlands, which, almost uninterrupted, extend for many hundred miles from north to south. (Sartorius 1961: 1)

On shore, Sartorius contemplates the evidence for landscape change along this coast. Other naturalists are making similar sorts of observations. By the time Sartorius has been published, Darwin is underway in the *Beagle,* with his copy of Lyell in his cabin. Sartorius observes that currents move sand into the mouths of the cross-lowland streams, such as the La Antigua River. The streams are impeded, they flood and lay down the sediments that keep the lowlands "inexhaustibly fertile" (Sartorius 1961: 3; 6.3). This does not yet take into account tectonic activity or the fluctuation of sea level, and Sartorius has compressed the time in which he imagines this to have occurred; for him it is an historic process. The full implications are still to be grasped, but the essentials of the process have been recognized.

Sartorius has winter north winds move beach sands southward from the diagonal coast, amassing and propelling "live" dunes. Vegetation gradually takes possession of earlier generations of dunes. Others tell about the ponds that form within the belt of dunes and point to them as sources of *mal aire* (figure 6.4).

In his text, Sartorius leaves Veracruz on horseback, westward out of the town and along the beach, which is littered with wrecks. This is the smoothest part of the journey inland. It is morning, the air is cool, and Sartorius sweeps the scene, fully engaging romantic German landscape description, complete with sandpipers, pelicans, crabs, and kingfishers. One can hear teachers in German schools treating their charges to this material in a story period at the end of the day.

Moving inland, countless previous hooves have loosened the sand and winds have scooped out the roads, making long depressions;

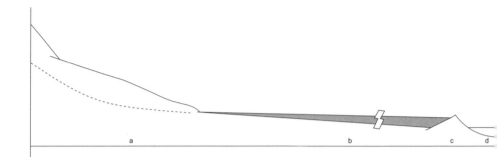

Fig. 6.3. Sartorius's hypothesis on sedimentation along the Central Veracruzan coast: (*a*) stream emerging from the hill land; (*b*) alluvium; (*c*) beach ridges; (*d*) the sea.

Fig. 6.4. East-west and north-south profiles of live and fixed dunes along the coast of Central Veracruz: (*a*) an east-west profile across the dunes with a thorny scrub vegetation (*chaparral*) and some "dry savanna" on the fixed dunes, as well as the beach, which was part of the road into the interior, and one of the wrecked ships that garnished the coast; (*b*) a north-south profile, parallel to the dominant winter winds, similar vegetation on the fixed dune, and hydrophytes in the water commonly ponded in front of dunes.

overarching vegetation turns them into tunnels. The air is stifling, wheels sink, and mounts struggle.

Sartorius lists the plants that fix the dunes, using terms that may or may not be currently traceable. Neither he nor any of the other authors mention fields or pastures on the sandy terrain. The soldiers of the American army need to get off the trails, to fill their canteens from the pools among the dunes and to fight through the vegetation. "A thicket—of bushes, shrubs, thorns, and vines so closely matted together that it is almost an utter impossibility for man or beast to pass through it" (Jamieson 1849: 25–26).

This has to be a fairly good approximation of what the early colonial authors called *arcabucos*. The soldiers call this vegetation *chaparral*: "The heat in this chaparral I lack language to describe; it radiated from the sands and danced about in front of you, impalpable but visible, like hideous phantoms of a diseased brain" (Kenly 1873: 326).

Probing farther now, into the bottomlands west of the dunes: the vicar, the médico, and the naval engineer, as well as various nineteenth-century travelers, have pointed in that direction, but here, finally, we are able to enter them, to have them described around us.

Sartorius chooses the right bank of the La Antigua River for a description of streamside vegetation that is often called, a little grandly, gallery forest. The naturalist goes into full flight, coopting the taxonomy of his time for a kind of poetry—enthusiastic, no doubt sensive, and largely Latin.

First the *Leguminosae,* especially the various species of mimosa—in graceful forms. Some other observers of the time took this term as a surrogate for all the vegetation between the bottomlands and the oaks near Jalapa. Sartorius comes to the mighty trunks of the fig and the *mamey,* just on the river's edge; some of them are still there. And then there is the giant grass; the vicar called it *zacat* (Hernández 1571: 195). A dictionary gives its scientific name as *Gynerium saccharoides* (Santamaría 1978: 204); it has various common names, including bamboo. It is used in the construction of houses and fences, it grows to thirty or forty feet high, and its extremities arch like ostrich feathers.

Across the river is La Antigua in the midst of lush vegetation, mirrored in the water: a tropical idyll, historically invested. This was the bustling town that had been reduced to a fishing village.

Sartorius strikes out now in a southwesterly direction; the exact route is unclear and does not matter. Again there is the fig, *Ficus americana,* an astonishing tree in these parts. It has a stout main trunk and horizontal branches, from which have descended shoots that become pillars, a vast, deeply shaded arbor. Vines, epiphytes, and gray veils of Spanish moss weave the space.

There are palms and more palms, a fascination to this naturalist and to those who have studied tropical vegetation in this region since, like Gómez-Pompa (1977) and Rzedowski (1978). These trees remain gracefully enigmatic, suspected to be indicators of human perturbation, as the botanists like to say, but it is not quite clear which species are diagnostic, nor of precisely what, nor of what epochs.

It is the dry season—it must be. Sartorius would not be in these parts in the wet season, when they would probably be flooded. His party is in the bottomlands, within "primeval" forest (1961: 8). They are on narrow paths, "often slippery and wet" (1961: 7). Occasionally there is a little settlement on higher ground, with freer air and some relief from insects. These, we may be sure, are the "islands" of older sediments within the wetlands. Within the forest too, one comes on herds of cattle.

Sartorius's readers will have wanted to know about wildlife, about tapirs and jaguars. They are there, he says, as are deer, armadillo, *aguti,* but all these are very seldom seen by the passing traveler. Snakes? Sooner or later the traveler who returns from tropical lowlands must say something about them. They are more frequent, he says, in the forest margins up toward the savannas.

He is soon up there and looks back toward the forest of the bottomlands, notes how it extends fingerlike up the valleys westward, and then surveys the sloping hill land itself, covered with brushwood and tall grass. Nearby is a "fine stone bridge" (1961: 8), and they are now quite near to Paso de Ovejas.

Other observers elaborate on the nature of the bottomlands. Manuel Domenech, press secretary of the Emperor Maximilian, describes a wetland along the new railway being built out from Veracruz. The Spanish prose is best left just as it is:

De este terreno, habitado por millares de caimanes, serpientes, sapos monstruosos y todos los anfibios que viven en terrenos fangosos, procede el vómito, esas exhalaciones pestilenciales que tantos estragos causa en los alrededores. La cuna del muerte está ataviada con la exuberancia de la vida tropical. . . . El agua estancada se oculta bajo nenúfares, plantas y flores acuáticas de grande belleza. Por entre breñas inextricables, enredades por lianas del famoso convolvulo de Jalapa, de flores de azur, del que los indios revelaron a Europa sus propriedades medicinales, se elevan palmeras, bananeros, palma christi gigantes, magnolios, lataneros (palmas brasileras), tulipanes caobas, árboles de cauchu y mil otra variedad de árboles y arbustos de forma elegante, entre extraño y eterno verdor. Colibríes de dorado plumaje, cotorras esmeraldas manchadas de rojo y gualda, garzas de largas piernas, e infinidad de lagartos y cuadrupedos pululan por estos desiertos mortíferos y feraces. Aquí se admiran mariposas de color azufrado revolteando graciosas entre tulipanes abiertos; por otro lado el nagazar atrapando artero a los sapos hidrópicos; más lejos rosas y flores de Angsoka abrigan bajo su hermoso manto a un podrido reptil sin vida y descompuesto por el calor. (Domenech 1922: 31–32)

This swirl of horror and delight converges on the key line: here is a cradle of death adorned with tropical exuberance. The outrageous plant and animal life, the very heat and rot, are fascinating.

Notorious Santa Ana, in and out of presidential office, had a favorite hacienda, Manga de Clavo, within the San Juan Basin. Rugendas has a painting of its yard and some unpretentious buildings with a view westward over the Basin toward Mount Orizaba (de la Torre 1985: 37). (The present town of Vargas is now on that site; the buildings of the hacienda are long gone.) One is drawn into the distant view as though examining the background of a medieval altarpiece. Green horizontal strokes lay out the bottomland, accentuated by some clumps of trees—hardly the forested terrain that Sartorius and others talk about.

Rugendas's most evocative image for these present purposes is one that appears with Sartorius's discussion of agriculture and ranching in the lowlands (1961: 180; figure 6.5). It is particularly attractive for its vegetation, which obviously engaged Rugendas's formidable draftsmanship, and which was given clarity, made vivid, by the process of engraving. The view must be northward over the wetlands northwest of Veracruz. The horizon fits, and we know that Rugendas roamed and sketched in these parts. Moreover, I have seen such places in the region. It is an "island" within wetlands, occupied by what seems a remnant of high forest. On the day I saw one, it was surrounded by the song of birds. Since then such islands and the birds that move between them have been engagingly considered by investigators at the Instituto de Ecología in Jalapa (Guevara et al. 1992).

Tree growth has no doubt been complicated by succession, but there are still several large specimens, a bit of what may have been a canopy, and several stories underneath laced with epiphytes, as well as the bushy growth that occurs when a high forest is opened to the light. Off to one side are the diagnostic palms, and around the tree growth of the island itself, as well as on slight rises all around, there is pasture of low grasses, grazed down in the dry season or even recently fired. Depressions in the foreground are filled with hydrophytes. This contrast is exactly what one sees on the remains of planting platforms and canals at this time of year.

Again, this is not quite Sartorius's "well-wooded plain" (1961: 7).

Fig. 6.5. Engraving of Moritz Rugendas's drawing of vegetation on an "island" within the San Juan Basin (Sartorius 1961: 180).

The two sources, interleaved, do not support each other. One is inclined to give greater weight to Sartorius, but one cannot be very confident. Historical deforestation and reforestation in and around the wetlands are just not very clear from documentary sources and may have been more ambiguous and variable anyway than is commonly recognized.

The nature and location of the neighboring ecosystem, the savanna, however, is another matter. We have been tantalized by the mention of savannas in Bernal Díaz, in the accounts of the vicar, the médico, and the naval engineer. The word *savanna* is written in on many of the sketched maps that accompany colonial litigation over land. For a moment it seems that this vegetation type is being distinguished on Corral's handsome maps, but no. Sartorius finally characterizes and locates it for us; some of his best passages are devoted to it. "One should not imagine pleasant meadows with wild flowers, but boring grassy plains overgrown with low thorny mimosas and studded by clumps of trees" (1859: 14; my translation).

There is a list of species too, which need not detain us. The characterization of seasonality is remarkable. The rains bring out the greens, Sartorius says. Thousands of cattle graze the rich new grass and enliven the plains. When the rains cease the grass grays, the trees lose their leaves, and the cattle seek out the canyons and the forests farther down. This is the time to burn, in order to cleanse the land of its vermin and encourage new growth.

In the earlier literature savannas are usually mentioned in association with hunting or game. So here, too, we are told again of the deer, coyotes, foxes, rabbits, wild turkeys, and lesser wild fowl. Sartorius goes beyond this to identify a stony but fertile surface (1869: 819). This is, or was a region amenable to agriculture. He notes scattered settlement, deals with the rancheros, to which we will need to return, and makes an excellent distinction.

It is not a simple matter, he warns, to say where the forest ends and the savanna begins; the first is found on fluvial deposits, the second on basaltic formations, and these are interdigitated. In just a few lines he has evoked a further, very schematicized profile (figure 6.6). It is useful for understanding the physiography of the western periphery of the San Juan Basin.

The sloping, savanna-covered piedmont, cut by canyons, is a "dull landscape. . . . Nevertheless this region has a peculiar charm for men of an enquiring turn" (1961: 10; figure 6.7). Sartorius finds it strewn with the remains of prehistoric settlement, water management, and agriculture; he sees them in the hill land all around his hacienda. He is especially interested in the fortifications that

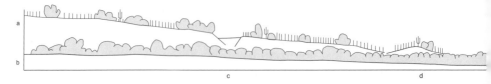

Fig. 6.6. Profile of the typical relationship between the eastern extremes of the piedmont and the beginning of the *barrancas:* (*a*) piedmont surface, often of a basaltic surficial geology, with savanna vegetation; (*b*) fluvial surfaces at the bottoms of *barrancas,* with forest (probably *selva mediana subperennifolia*); (*c*) a converging *barranca;* (*d*) an outlier of Tertiary sediments.

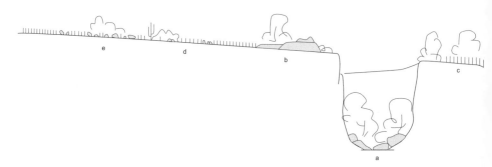

Fig. 6.7. Profile of piedmont surface littered with Prehispanic remains and a *barranca:* (*a*) stream bed among rocks fallen from the cliffs, with forest vegetation—sometimes ancient dams can be detected; (*b*) an archaeological site on a headland; (*c*) savanna, the predominant vegetation of the piedmont surface; (*d*) surface rocks gathered into stone lines; (*e*) the natural surface rubble, including sizable volcanic ejecta.

dominate some of the barrancas, and in the remaining monumental architecture (1869). He finds that some of it has already been torn down to build the bridge over the La Antigua River, at a place that came to be known as Puente Nacional.

Various colleagues and I have followed up these observations on air photographs and in the field. We have been able to verify some of Sartorius's observations but have also been impressed with very extensive webs of "stone lines," which we have interpreted as the bounds of ancient fields cleared of some of their stones—an intensification of rainfed agriculture, meshed with intensive wetland agriculture downslope (Siemens 1989c; Sluyter and Siemens 1992).

The evidence for assiduous ancient land and water management, for considerable population concentrations in an area only very sparsely populated now and used mainly for ranching, makes Sartorius reflect on what happened. The achievements of ancient peoples in this region contrast dramatically with what Sartorius sees is being done by those who live here now and thus pose another invidious comparison, grist for the denigration of the lowlanders, particularly the Indians. Also, if ancient people were able to use the land and water of this region effectively, then surely colonists of good German background will be able to do the same again, and more.

The sloping plain of the piedmont is "rent by fearful chasms" (1961: 9). Sartorius explored quite a few of them and enjoyed it. He characterizes cliffs, caves, streams, waterfalls, and vapor-nourished vegetation with romantic verve and laces the description with science—as naturalists of his time tended to do. He observes the stratigraphy and considers the forces responsible, particularly volcanism and stream erosion, as well as the enriching effect of the running water. In the rainy season this is strong enough to move boulders. Where canyons are wide enough, residual moisture in alluvial soils nourishes pasture or provides agricultural opportunities. Indians have fields and gardens there, in places that seem quite inaccessible. One is reminded of the *barrancas* evoked in the sixteenth-century materials (figure 4.3) and what one can see to this day when one flies over or descends into them.

Movement over the piedmont is characteristically east-west; any attempt to move north-south is very laborious. The narrow canyons can be crossed on felled trees, along precarious suspension bridges, or by means of various other contrivances of lines and baskets (figure 6.8). In wider canyons one descends on trails and then is ferried over the river at the bottom by Indians on rafts of bamboo staves and gourds. It is great fun, Sartorius tells us, to take visitors down on mules and frighten them. But one must watch how one goes: vegetation can obscure precipices and there is falling rock; in the rainy season trickles can become raging torrents; and one should not leave the return ascent too late in the afternoon—darkness descends quickly.

The Lowlanders and Their Work

Sartorius's lowlanders are not as impressive as the lowlands. This is now the darker side of Humboldt's dyadic pronouncement, embellished. However, if one can hold one's nose for a bit, a fairly comprehensive human geography emerges, more specifically an agricultural economy and ecology of the margins of wetlands and neighboring hill lands, together with an outline of ranching within

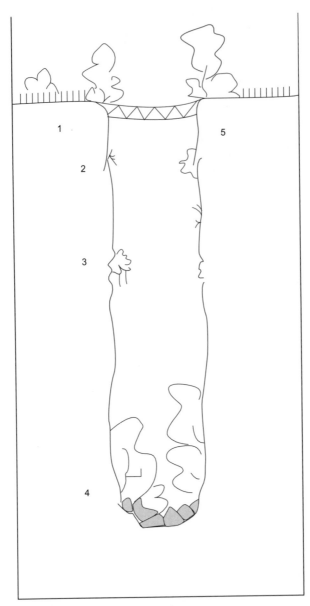

Fig. 6.8. Profile of a narrow *barranca*.

the bottomlands. Moreover, one detects a basal congeries of sustaining activities, sketched out inadvertently. This can be projected both forward and backward in time; it enlivens the spare references to livelihoods in the colonial sources and informs agrarian development in the lowlands into our time.

Sartorius's representation of lowlanders and what they do must be synthesized, like the landscapes, from scattered comments. His terms for rural people often overlap, his exposition can be labyrinthine and his style antique, but his material is rich. He does some quite masterful things with diagnostic verbal pictures—vignettes in the romantic manner, complete with what have to be constructed conversations. Behind all this, we can assume, are the encounters of many years.

The characterization of the lowlanders begins with a vignette on the wharf at Veracruz: an array of the curious, watching foreigners arrive. Various travelers rendered that tableau, as anxious as Sartorius to begin their characterization of Mexicans. In the vignette, Sartorius sets the categories that will organize much of his discussion of Mexican people: blacks, mulattos, mestizos, and creoles— the Indians are elsewhere and he soon gets to them. For the moment he is distracted by the women in their loose clothing; that happened to other observers too.

The Indians ghost through the peripheries of the landscape seen from old Veracruz; they are scarce and peripheral still, here in the middle of the nineteenth century. In Sartorius's prose they are associated with the canyons. Indians are the best guides there; they clamber like goats. Sartorius maintains that they like the danger and the solitude. They overnight in caves, and they do not fear the prowling jaguar or the swarms of monkeys. They have their fields there and practice their heathen rites in the *barrancas;* Sartorius has seen the evidence.

From our perspective, considering the various historical materials that are available to us, Sartorius's description of Indian life in the canyons is particularly intriguing in that it shows effective adaptation. These hidden places offered refuges from pursuit and oppression; they provided spiritual retreats, and their various microenvironments could be made to sustain. We owe an extensive consideration of regions of refuge, in general terms, to Mexican anthropologist Gonzalo Aguirre Beltrán (1991). There is more to be done regarding the nature of the pressures in this region and, indeed, the history of the occupance of the *barrancas.*

In the sixteenth century the production of food was associated with Indians; the rest of the population was interested in other things. The condescension toward the Indian was condescension toward agriculture. In the world according to Sartorius, subsistence agriculture is still largely Indian; it is of greater direct interest, perhaps, but it is still depreciated.

Sartorius explicitly denigrates the Indians in a variety of ways. They are suspicious, taciturn, and ambiguous when they speak. "They want the broad and lofty forehead. . . . We find in them the talent of imitation and comparison, perhaps humor and wit, but no poetry"(1961: 64). Eventually Sartorius gets to a main point: They are "without energy, brutalized by intoxication, and confined to the narrow circle of a stereotype existence and train of thought. There is consequently, no reason for apprehension" (1961: 66). These Indians will not contest colonization.

In Rugendas's print of the Indian family in front of their house in Jalcomulco, a town in one of the *barrancas,* condescension is visualized (figure 6.9). The house is rustic, of course; it is surrounded by fine vegetation, which the process of engraving renders sharply and with a certain grace. Domestic animals are among and around the people, which seemed promiscuous to more than one observer. Everyone is in costume; the women's clothing just loose enough to be intriguing. A line of men bearing loads passes by in the background. One is reminded of tableaux in the Museum of Anthropology in Mexico City. They are quite as charming and as condescending as this nineteenth-century antecedent.

At the opposite end of Sartorius's spectrum are the creoles: natives of Iberian descent and mostly townspeople (1961: 53–61). They include government officials, higher clergy, professionals, merchants, and the owners of factories, mines, and large landholdings. It would have been very useful to have been able to follow up on some prominent colonial Central Veracruzan names and properties, but Sartorius describes only the class. The towns are not dealt with specifically either, for that matter, except in regard to the sluggish trade in the port of Veracruz—both lacunae are diagnostic of the author's own personal inclinations.

Black people are not very numerous in Mexico or even in Veracruz; they are diminishing as a separate element, he says, their characteristics diffusing through miscegenation (1961: 50–51, 82). Black slavery is a thing of the past, in the country and in the state. Another author subsequently agrees; it had happened fairly gradually around Jalapa and violently around Cordoba (Carroll 1991: 93–111). The numbers of black people in the port towns is being kept up temporarily by migrations of craftspeople from Cuba and the West Indies—no doubt an interesting story if it were to be followed up!

Fig. 6.9. Engraving of Rugendas's drawing of Indians against their house in Xalcomulco (Sartorius 1961: 64).

The black people and those of mixed races are commonly grouped together into "castes" and distinguished from whites and Indians—those are the three official categories. Sartorius pays careful attention to the people of mixed races, taking us into some cultural and, indeed, overlapping designations.

Jarocho is the general term used for racially mixed people of the Veracruzan lowlands. It was originally a disparaging term that uplanders used for lowlanders and then subsequently became a term that lowlanders applied to themselves with pride. It is versatile: an association with herding and horsemanship is common, but it can be used for agriculturalists and fishermen too; in fact, it can be used for urban as well as rural people (Siemens 1990: 158)—a virtual synonym for *Veracruzano*.

The jarochos that most interest Sartorius are the rancheros; their work and their ways of making do both attract and repel him. They have been much discussed of late (e.g., Skerritt 1993); a dated characterization by a careful observer is offered here for whatever that may contribute to the understanding of this figure.

The ranchero is definitely not Indian; he has little in common with the stolid people relegated to inconsequential agriculture. He is shown as incomparably dynamic. He is not an hacendado either, all the usual trappings of that term do not apply: vast holdings, owned but heavily mortgaged, absenteeism, the house in the town. The ranchero's holding is of a smaller order of magnitude. It may be owned, but rental of land for agriculture and a fee for pasturing one's animals seems to have been more common. There is a range in the well-being of the rancheros. Some content themselves with a few animals and the insubstantial agriculture that foreign observers in these parts, not only Sartorius, disdain. Others are better off, with sizable herds or fields of sugarcane, which they process in their own rudimentary mills, called *trapiches,* and sell in the form of a crude brown sugar called *panela,* which provides them with cash income.

The rancheros, we are told, may live in villages or towns, but mostly they live in widely scattered dwellings (ranchos), or in loose groupings of single-family dwellings called *rancherías.* Sartorius makes a few good points about this somewhat enigmatic rural settlement form—garbled in all the translations, as it happens (1859: 319). The *ranchería* may consist of a scattering of residences, but the rancheros living in it pertain to a community that elects its justice of the peace and its police; they have a sense of community, an *esprit de corps,* the former member of a German student *Burschenschaft* is led to add; when there has been cattle rustling, for example, they are quick to organize a posse.

Skerritt has pointed out that rancheros still tend strongly toward independence and share a set of values that have issued from cattle management and horsemanship (1993: 12–14). That is already clear from Sartorius, but he takes it further:

> The Mexican is fond of cattle-breeding, because it feeds him without hard work, enables him to indulge his taste for a Bedouin life, and to be on horseback as often and as long as he pleases. This is why, in addition to the great cattle-breeders, there are so many *rancheros* who carry on cattle-breeding on small farms [the German actually says, 'rented land']. (1961: 181; 1859: 310)

There is the *vaquero,* the employee on the hacienda or the larger rancho, to speak of still, as well as the whole fascinating business of ranching. First, however, Sartorius must be allowed to explain— mostly to putative colonists, it would seem—what can be grown in coastal Central Veracruz.

Sartorius's Sketch of "Agriculture in the Torrid Regions"

Subsistence and potential commercial crops can be grown with the moisture of the annual rainy season. Perennials, too, can be grown without irrigation. So, Sartorius says, water will be no problem. It is curious, or perhaps not, that he should have passed over several lessons here. Central Veracruz is in large part a subhumid enclave within the lowlands. Water was gathered and conserved in Prehispanic times, as Sartorius found to his own great amazement, presumably to facilitate agricultural intensification. Water could have become a problem indeed in any sizable attempt to colonize and farm commercially, but that is probably not a point that a promoter of colonization would want to make.

Sartorius's discussion of potential commercial crops is at first a catalogue of problems. Sugarcane, the most important commercial crop of the terrain below about 1000 metres during the colonial period, has been severely prejudiced during the wars of independence. In the plantings that remain, South Sea Islands varieties are common; they require eighteen months to ripen for the first time and then can produce two or three more harvests annually. Most of the best plantations were in the hands of Spaniards and hence targets during the conflict. In the surroundings of Cordoba alone, thirty-six sugar haciendas were destroyed. Whatever is left for a year in this climate becomes a forest. At the time of writing, he says, scarcely enough sugarcane is being grown in Mexico to satisfy home consumption; prices of raw sugar in Mexico are as high as those of refined sugar in Europe.

Cane was the mainstay of El Mirador for many years (Pferde-kampf 1958: 163). The ruins of Sartorius's processing plant can still be seen across the road from the main buildings of the hacienda. It used an imported steam engine and a mill made in New York; its main product was a crude rum called *aguardiente.* The chimney is still intact, its rim decorated with some fairly delicate brickwork—a whiff of antique sensitivity among the trees.

Cacao is grown in the hotter, moister, lower parts of the lowlands, i.e., in Tabasco, not in Central Veracruz. As he explains this, Sartorius makes very clear that these are the tropics that he considers impossible for any but acclimatized natives; he advocates new settlement in the temperate tropics.

Vanilla is harvested in coastal Veracruz, but mainly to the north, in the homelands of the Totonaca Indians. The description of its exploitation is a gem (1961: 176–177). This plant is a vine that grows in the forest but must nevertheless be tended; its pods require very careful handling. In a country where labor is scarce, this cannot be done on a large scale. It is the work of Indian families—an inadvertent tribute to the old ways.

Tobacco could be hugely profitable, since everybody in Mexico smokes, we are told, and foreign markets are ravenous for it, as well. However, the sale of tobacco is a governmental monopoly—it was so during the colonial period and remained so after independence. Production and transport are rigidly controlled; illegal trade is difficult. Sartorius must draw the lesson. There is just enough for internal consumption, whereas it could be an article of profitable export. Look at Cuba: it has more liberal trade laws and its ports are forested with masts.

Tobacco could do particularly well in Mexico. It can be widely grown within the annual rainy season: the plants are grown in hotbeds in June and July, transplanted into the fields from August till November, and harvested, several times, from December to April. That suits the Mexican agriculturalist just fine; he likes to work for part of the year only.

Sartorius goes on to describe a series of crops with more realistic potential for the sustenance and gain of the envisaged colonists.

These crops are grown with seasonal moisture in the context of horticulture and shifting cultivation on the holdings of the rancheros.

Coffee is a promising new introduction, not yet exported. It does best around 1000 metres; many rancheros in the upper piedmont grow it in gardens around their dwellings, under orange, mango, and banana trees. The berries ripen gradually from November to March, he says; they can be harvested leisurely; picking, cleaning, and drying is the work of women and children. "For the small planter it is just the thing" (1961: 175).

Rice, Sartorius's readers will perhaps have been surprised to find, is planted widely in coastal Veracruz by the rancheros and the Indians. It is not swamp rice but upland rice, grown in the context of shifting cultivation. It yields an abundant harvest and rarely fails; it, too, could someday be an export.

Cotton is interplanted with maize. In the coastal Veracruz of the nineteenth century it is an annual and a shrub; the vicar's cotton is perennial tree cotton. Both require a strongly seasonal climatic regime: high temperatures, a substantial rainy season, and winter months without rain (Sluyter 1990: 56– 61). Along the coast it does best below about 1000 metres.

Maize is the mainstay, of course; all the rancheros plant it, plus beans, chiles, tomatoes, and various tuberous plants. Sartorius mentions arum root, manioc, sweet potato, and yam, which richly provide the starchy foods required and obviate the potato, he tells his German readers. All of these food plants are in the scruffy little fields around the dwellings. The ranchero takes little trouble with them, our author maintains, considering them as snack foods. "If he cultivates half an acre of them, [he] thinks he has done much" (Sartorius 1961: 179).

Such perceptions surface in other foreign observers' accounts, as well (Siemens 1990: 150–152). The small plantings are hard for them to dignify as agriculture. Also, accustomed to eat more formally at table, they do not understand, or perhaps just do not approve, of the apparently haphazard way in which the natives gather their food, cook it, and eat it.

And then there is the scandal of the banana. In its second year this plant already has a fruit branch; for half a century it will continue to produce fresh branches, as long as any intruding brushwood and the branches that have borne are cut away. The fruit is in various degrees of development on the various branches; some of it is always ripe. It is eaten raw or cooked and is prepared in all sorts of ways; its juice can be fermented and that liquor can be distilled.

The foliage provides a pleasant shade, the dried fiber of the stem can be made into a cushion, and the huge leaf can be used as a table cloth. This plant, more than any other, makes life in these lowlands too easy.

Piqued by the banana, Sartorius pauses to explain what is wrong with lowland agriculture. It is ironic that a beneficent plant should bring this on—in the mind of a botanist—and indicative of the strength of the conviction that the lowlanders are lazy. But the passage, which is only in the German original [1859, p. 305–306], also indicates that Sartorius is not so out of touch with the realities of his time as von Mentz has deduced (von Mentz 1990: 12).

Evidently there was a proposal in Spanish times to eradicate the banana in order to force the lowlanders to become more active. Sartorius agrees that indolence was then, as it is in his time, a basic deterrent to progress, but notes that there are other reasons for stagnation than an overbounteous nature. The Spanish should have taken care to raise intellectual and moral levels—here is his liberalism fully to the fore. They should have seen to education and regional policing, improved the means of transportation, built roads and canals, assured the possibility of the private ownership of small parcels, and protected agriculture against ranching. In his time it dominates the economy and the landscape disproportionately, constraining the agriculturalist to spend all his profits and much of his time in fencing.

Granted, much of the nineteenth century is Mexico's *epocha negra* and stronger things are being written about the economy of the country (Siemens 1990: 38–39, 162–163). Undoubtedly Central Veracruz is affected by the conditions outlined. External trade, until the final quarter of the century, is very sluggish; there are usually few ships in the harbor of Veracruz, certainly in comparison with Havana. They normally carry out only low-volume, high-value products: gold, silver, cochineal, vanilla, a few drugs, and skins. They must go to Campeche to get logwood or load up in Cuba for the return voyage. Internally, there has been much destruction in the cities and villages during the various military campaigns, and the main road is a ruin that will not be repaired for many years. High prices, arbitrary government regulations, corrupt officials, thievery, limited markets locally and none farther afield, and lack of currency, labor, and credit—all this makes it very difficult to do business of any kind. Land rents are high and contracts of short duration. There is a great deal of rustling; horses and cattle are often confiscated by soldiers. The very poor means of transportation disarticulate the economy of the countryside; a surplus in any given locality is a disaster. A Mexican author asks, Who would want to work when there is no gain? (Iglesias 1831 [1966]: 22). Under these circumstances indolence is adaptive.

It is difficult to imagine that Sartorius was not aware of all of this. It is likely that he was evading much of it in print because of his desire to promote colonization.

In what he does say about rural conditions there is an intriguing ambivalence. He is frustrated; the agriculture he sees is insubstantial and disorderly, there is no sign of the plow anywhere. It could all be done so much more rationally and energetically—but there is something else: "[The jarochos] are not fond of hard work; nor have they any need of it, as they have plenty to live upon" (Sartorius 1961: 181). Disapproval repeatedly shades off into something close to admiration, as will be seen. This just survives translation; it is clearest in the tone of the original.

Further, as one works through the descriptions of life in the lowlands, the sparse commentary on problems and the much more elaborate description of potential, the agricultural and horticultural matrix treated in passing becomes more impressive than the featured crops. Sartorius can hardly have intended it, but that is the effect.

This subtext begins early in the book, when Sartorius first takes us through the bottomlands behind the port. Whether it is a construct or an account of an actual visit, he asks for refreshment at a rancho along the La Antigua River, between the sea and the village. He lists what is available around the hut: bananas, pineapples, oranges, avocados, and a variety of other fruits. Palm-wine has been made, beans are simmering on the fire in the middle of the hut, and the stones for grinding maize are there just to the side of it. In quick strokes he describes the building, too: beams, rafters, and laths fastened with vines and thatched with palm leaves; walls, shutters, and doors of bamboo staves (he undoubtedly means the substantial stalks of cane grass). Such a house needs no windows, since the slits let in the light—an amazing indecency to various foreign observers, some of whom called these chicken coops (e.g., Mayer 1844: 10). The furnishings are very simple: benches, bedsteads, and a shelf for the utensils, also of bamboo. "How little suffices for these natives of the tropics" (Sartorius 1961: 5).

Reformatting scattered comments on the rural economy allows us to enrich this rendition: Most rancheros keep some cattle, which they pasture against a fee on private lands or simply allow to forage where they can. These animals are something to fall back on when extraordinary expenditures are required.

Various other animals surround the ranchero's household: donkeys provide transportation and dogs keep watch. Pigs are fed on scraps, and household fowl range in and around the house—all important sources of fresh food and something to set before a visitor or to sell when necessary.

Anyone living within range of fresh water has access to turtles and fish. Sartorius does not make much of them, but one recalls the vicar's appreciative descriptions; they will all still have been available, as well as game: deer, wild boars, and wild turkeys.

Agriculture requires only a few hours of work a day during the rainy season and then none at all for half the year, at least so it seems to Sartorius and he may well be underestimating the effort and care involved. During all that off time, ready money can be made in various ways. Charcoal can be prepared and sold in Veracruz. Cordage can be made out of the fibers of the maguey and taken to markets or one may collect various salable materials out of the forest, depending on just where one lives. The men can hunt and then tan and dye hides for their own clothes; the women can spin and weave cotton. Fuel and the materials needed for tools, furniture, and other household necessities are close at hand.

All this can be bracketed by another, more elaborate verbal sketch, near the end of the book. Sartorius stops again at a rancho to rest and ask directions. He observes the domestic animals, looks around him at the interior of the house, as before, and exchanges pleasantries with the lady of the house while they wait for the men to return from their work. The man and his four sons arrive, all in coarse blouses and sandals, and lay down their loads.

In the talk that follows, the man of the house explains something of the altitudinality of their situation. They live on a high point for the breezes and relative freedom from insects. One cannot cultivate in the immediate surroundings because cattle roaming the savannas would give one no rest day or night, breaking through hedges and ruining crops. Fires sweep over this land, as well; a cleared circumference is necessary to protect the dwelling.

The crops are planted down in a canyon. There is water and excellent soil, with residual moisture well into the dry season, no doubt. Steep topography and a perimeter of bush keep the cattle away. One cannot live down there because of the fever one would catch, the snakes, the mosquitoes and ticks. And one could not keep fowl or pigs down there either; they would be taken by foxes, coyotes, or one or another of the larger cats.

The loads that the men have brought sum up their way of making a living: a supply of maize, some firewood, a jar of water, a basket with manioc and fruits, and a living armadillo. Slipping into a textbook mode, for a moment, one may observe that the jarochos cultivate, but they also hunt and gather. They combine a wide range of activities in a suite of adjacent microenvironments into an effective, flexible combination: a subsistence system, producing what one needs for immediate consumption and the means to obtain what must be traded in.

Altogether, one can also deduce a rather good diet. Beef is plentiful and inexpensive, and a whole range of additional animal protein is easily available. Staples are readily produced; sweeteners, flavorings, and other amenity foods are abundant.

There is one particularly telling passage. When the jarocho, and we can infer ranchero, needs water, he ties large empty jars over a donkey's back and rides with them into a stream. The jars fill of themselves and he takes them away. When he needs fuel, he cuts down a tree, pushes it through the door to his fire, and progressively feeds it in until all is consumed, without having had to cut it up. "That is tropical *savoir faire*" (Sartorius 1961: 6).

Most rancheros, he tells us, cannot read or write; they have had no schooling. Before they marry they must know some of the catechism off by heart; that is the only religion they learn. Many days they lie about playing the guitar or staring at the sky. Should there be news of a *fandango*, a party, they bathe, anoint themselves, and become indefatigable in song and dance.

A reflective passage follows—not translated. As careless, indolent, uneducated, and undisciplined as the scattered rancheros of Central Veracruz may be, they are not vicious; they share what they have and can be trusted. And compared to the peasants of the heaths and mountains of Sartorius's homeland, they are happy and free of care (Sartorius 1859: 309).

Ranching

Sartorius is not a cattleman himself, but he habitually travels on horseback, appreciates the horsemanship of the lowlanders, and is obviously fascinated by what happens on ranches. His observations interlace with those in several recent sources.

José Luis Melgarejo Vivanco's *Historia de la Ganaderia en México* (1981) is one. The exposition is undisciplined, the chronology cavalier, and the text unencumbered with the conventions of scholarship, but there is obviously much experience here, as well; the author is a *Veracruzano* and a *ganadero*, among other things, and often quite incisive.

Fortunately there is also David Skerritt Gardner's *Una Historia Agraria en el Centro de Veracruz: 1850–1940* (1989a), which is a different matter. Its data base is detailed archival material on properties, population, and change in a swath of territory that begins up in the hill land not far short of Huatusco and ends near the sea around La Antigua, cutting across the upper San Juan Basin. In the background is the basic work on Spanish American ranching by various scholars, including Butzer (e.g., 1988), Doolittle (e.g., 1987) and Jordan (e.g., 1993).

Sartorius's lowland Central Veracruz is still very sparsely settled: "Many hundred square miles of the most fertile soil would be completely useless to the proprietors, if they were not employed for pasturage, which can be attended to by a few herdsmen" (Sartorius 1961: 181).

The focus is on the *hacienda del ganado*, a large estate with solid buildings. And immediately there is an interesting distinction. On the plateau the *cascos* of the haciendas look like castles, with high walls, battlements, and turrets for defense. Windows are barred and the main entrances are closed for the night with heavy crossbeams. This is not common below the summit; the owners are not so concerned about defense, Sartorius maintains, and this seems indeed to have been the case, judging from Rugendas's painting of Manga de Clavo (de la Torre 1985: 37), our own search for colonial remains in Central Veracruz, and the recent Cambrezy and Lascuráin (1992) study of the haciendas of Central Veracruz.

The *hacienda del ganado* typically encloses a vast extent of land. Between the upper piedmont of Veracruz and the sea, there was still, early in the nineteenth century, a truly massive holding: San José Acazónica (Skerritt 1989a: 65–67; Pferdekampf 1958: 159). Such properties are described with a sweep of the arm. The westward margins of this one lay between Jalapa and Cordoba, from where it extended in a broad band to the coast. It measured 120 square miles. These were Jesuit lands in the colonial period; at the time of Sartorius's arrival they were held by Francisco Arrillaga. Sartorius purchased the land for El Mirador from him. The gradual subdivision of large holdings was going on elsewhere in the region, too.

It becomes more and more difficult to separate hacienda from rancho as the nineteenth century wears on. Skerritt reminds us that the hacienda is a socioeconomic entity, with a particular set of relationships between owners, administrators, and permanent workers (some resident on the property, some not), plus occasional workers, renters, sharecroppers, and partners in specific ventures. The rancho, in contrast, is operated by a family, with occasional additional labor drawn in (1989a: 70–71). Sartorius stresses that it is a smaller holding, owned or rented, and devoted to agriculture as well as ranching on a small scale (1859: 274). In his time the *hacienda del ganado*, devoted mainly to cattle-breeding, is still an important entity. On such properties within the coastal region between 10,000 and 20,000 cattle might be pastured. Santa Ana, the great scoundrel of Mexico's *epocha negra*, was a contemporary of Sartorius; his holdings were of this order of magnitude (Calcott 1936).

The owner may or may not live on his holding. Some spend a considerable part of each year in an urban residence. One of Sartorius's best verbal sketches is a large, lumbering coach of such a household in transit, tied about with chests and bedding, pulled by half a dozen mules, and surrounded by a swarm of mounted servants (1859: 275). He is even better in his *costumbrista* profile of the owner who lives on his estate all year—preferring the rural life, as did Sartorius himself. Such an owner takes a newspaper and knows what is being discussed in Congress; the priest comes to pay his respects every Sunday, and the civic authorities will also visit from time to time. The hacendado inspects his property on a fine horse and in a saddle trimmed with silver. The cut of his clothes is as the ranchero's, but of a finer cloth; he wears cologne but cannot entirely mask the smell of the stable (1859: 275). "The owner of such an hacienda lives like Laertes or Odysseus on Ithaca" (Sartorius 1859: 319; my translation). He can be authoritative, engaged in actual work as well as supervision. He is bathed and invigorated—from time to time, by young women—and dines in "goodly halls"(Homer, in Butcher and Lang edition 1965: 280–283). The allusion was quite possibly rueful for the owner of El Mirador. His people had not proved quite as tractable as the thralls on Ithaca.

On the *hacienda del ganado* there is always a chief vaquero, called a *mayoral* or *caporal*, who is magnificently knowledgeable about all aspects of ranching and is a living chronicle of the region. Sartorius finds him a joy to hear, and tucked into the original recollection of corral-side talk but left out of the English and Spanish translations, there is an expression of frustration (1859: 316). The nuance of the disquisition of the *mayoral* cannot be purveyed in German prose. This is reassuring, an indication of Sartorius's own abilities in Spanish and hence the incisiveness of his ethnography—not his judgments on the people of the lowlands necessarily, but his appreciation of what was being said and done.

A given owner's total herd is divided into units, each one both a number of animals and a place. Sartorius has looked at this carefully; he seems to find the animals themselves as responsible for this as those that herd them. They keep together, "in herds or families, and choose favorite spots, to which they return"(1961: 182). These seem to be the quadrants of an operation. They are elusive units, designated singly or in some further combination as *hatos, sitios, vaquerías,* or even *estancias.* They are part of the vagueness, the "extensive" nature, of animal husbandry in premodern Spanish America.

The five to eight hundred head of a basic unit are under the care of a vaquero, who lives at a watering place out in the middle of the habitual pastures of this herd, adjacent to a corral of stones or logs. Sartorius admires him, frankly. He is trustworthy, he keeps good tallies of births and deaths on leather straps; his countrymen look up to him, and he rides very well, flat down on the neck of his mount and fast under low trees, accurate still with his *lazo.* But then good horsemanship is general in this country, reprehensible perhaps for the lack of industry and sobriety that it implies, but irresistible just the same.

Horses are a small minority among the animals on coastal holdings, Sartorius maintains, and this accords with Corral's figures. They are needed for cattle management, as mules are needed for freighting, but both are best bred on the uplands. Sheep and goats do not do very well in the lowlands, either. Thorny vegetation ruins the wool; foot rot and other diseases carry the animals off in the rainy season.

Bovine cattle are the main thing. The best profit is the sale of oxen and old cows to the butchers. The market is strong, since everyone, even the laborer, is accustomed to eat meat everyday; it is the cheapest food. Few haciendas have dairies; such profit as might arise from milk is left to the vaquero in order to induce his attention to the taming of the calves.

Are there still *cimarrones* in 1850? Most cattle are marked, on their ears as well as by brands, and they respond to the attraction of salt. However, on the large ranches that are not able to engage enough vaqueros, many cattle escape and become feral. They flee at the approach of people and cannot be rounded up unless one uses tame oxen called *cabestros* as decoys. Sartorius indicates clearly that this represents inadequate management; he does not romanticize *cimarrones.*

The tame and the feral range freely most of the time, like deer in a park. But whether driven or left to their own devices, they move seasonally: "During the rainy season [they are] in the savannahs, during the dry months in the shady forests" (Sartorius 1961: 182).

Here, finally, is a simple articulation of the local transhumance that is inferable from the early colonial years onward. It encompasses both the east-west movement from wetland to hill land and the vertical movement between the sloping surface of the piedmont and the *barrancas*. It remains a metaphor for interdigitated agricultural use of neighboring microenvironments from the distant past to the present time.

7. Struggling with a Technocratic Pathology of the Basin in Mid-Twentieth Century

"Un considerable disaprovechamiento."

—Sergio Morales Rodríguez, in SRH 1976a: iii

Highway 140 from Jalapa to Veracruz, as it angles through the southwestern extremity of the Basin, passes various small communities and then, just east of Puente Jula, crosses a large, choked drainage canal with ridges of grown-over spoil on both sides. We would come this way on many mornings during our investigation of the remains of Prehispanic wetland agriculture in the bottomlands to the north. The road that goes off from Highway 140 in that direction, toward Vargas, runs alongside a dry irrigation canal, its sides breached in many places, its floodgates ruined. While we were investigating one intervention, we were being reminded of another.

Technically trained people in the Secretaría de Recursos Hidráulicos (SRH), the federal governmental agency charged with the development of river basins and thus responsible for the sad relicts, bring our discussion of the San Juan Basin and its favor into recent times. They are convinced of the direct applicability of science and of the imperative of agricultural intensification by irrigation. Much of the information they marshal in their reports has already been useful in the characterization of the natural environment of the Basin, in chapter 2, and in the clarification of historical materials. Here it is used to sketch out a further geography of favor. The technicians treat human agency shabbily. Fortunately there are other recent data and personal observations that can be drawn in.

Those who write the reports give the Basin a negative cast—any reference to favor is grudging or implicit. This is made to serve, nevertheless, allowing the analysis one further representation of the Basin and a trace of various continuities of land use into the recent past.

Taking a Good Deal of Agrarian History as Read

A move from Sartorius's Central Veracruz in the mid-nineteenth century to the modern engineers' San Juan Basin of one hundred years later again vaults many years. Much of what affected the Basin in the interim is fairly easily accessible in various thoughtful publications (e.g., Brading 1985; Fowler 1979). The Veracruzan historical studies by David Skerritt Gardner (1989a, 1989b, 1993a, 1993b) identify a series of particularly interesting themes. Modernization is already detectable in the mid-nineteenth century; on Sartorius's hacienda, El Mirador, new machines and techniques are being introduced and the promising new crop, coffee, is being grown. The process ripples through the economy of the region—not nearly as rapidly as various observers would have liked to see, but it does gather momentum. Skerritt describes the progressive hacienda El Faisán, located in the northwest quadrant of the Basin, west of La Antigua. Before the end of the nineteenth century it is producing beef and milk for the market in Veracruz city, with animals of European breeds, nurtured by African grasses.

The traditional hacienda is giving way, large old holdings are being subdivided, and the rural work force is becoming more mobile. The rancho, a family production unit, becomes more and more noticeable, and the ranchero becomes an intriguing figure. "Radical" times set in after the turn of the century, of course, with peasant and worker movements, in-migration, affectations of large holdings, ejidalization—and special exemptions from affectation. The political history of Central Veracruz becomes quite fascinating in the 1930s and 1940s; people who will become important figures on state and national levels emerge. All the while certain prosaic basics persist: commercial sugar production, cattle ranching, and what Skerritt Gardner has called an "agricultura básica" (1989a: 14). In the Basin something new and, at first sight, quite dramatic is being undertaken.

The Mid-Twentieth-Century Hydraulic Intervention

On a wall of the Las Palmas restaurant in Cardel there is a photograph taken from the inside of a tunnel, silhouetting two men in a contemplative attitude against the round entrance. At the bottom there is only the date: 1936. The tunnel seems to belong to a system built on the north side of the La Antigua River in the late 1930s by the Comisión Nacional de Irrigación, the antecessor of the SRH. The system included a take-off dam, a trunk irrigation canal via a

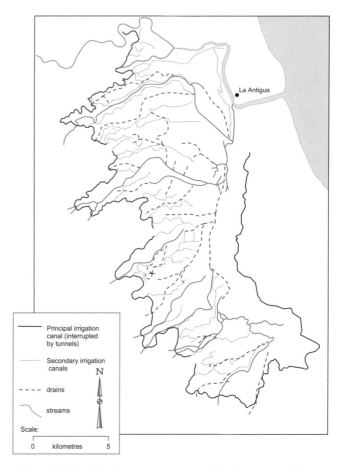

Fig. 7.1. Map of the hydraulic intervention of the Secretaría de Recursos Hidráulicos in the San Juan Basin (Comisión Nacional del Agua/SARH, Plano General Catastro, Distrito de Riego No. 035 La Antigua, 1987).

tunnel through a spur of hill land, distributory irrigation canals, drainage canals, and the necessary roads; it made about 2500 hectares irrigable. The system was soon damaged by the river during one of its seasonal highs and rendered useless. In 1947 SRH studied the district again and then undertook a reconstruction and an elaboration southward (figure 7.1).

During the remaining years of President Miguel Alemán's *sexenio* (1946–1952) the dam was rebuilt, a new canal was constructed around the western and southern sides of the Basin, and the Basin's rivers were "rectified." Older residents remember the SRH crews moving through the countryside and the massive steam shovels against the horizon, cleaning out, deepening, and straightening the channels for a better flow.

The initiatives of the Alemán period left certain fundamental problems unsolved; they are bothersome to this day. The most serious is the exit of the system; that is, the entrance of the San Juan into the La Antigua River. It is not low enough to drain the Basin entirely—colonial observers had already noticed that the bed of the first was lower than that of the second at the junction (figure 4.1). Another problem is that the tunnels southward were constructed without facilities for flushing out sediments. Men and animals have had to go in at intervals to clean them out; machines cannot be used because of inadequate ventilation. Strange things are found on occasion, including human remains.

There is a monument near the river, memorializing the inauguration of all this in 1950. (It was more or less in operation by 1952.) Right under the heading, the inscription sets out some statistics, and the first of these is the pharaonic figure, the number of hectares considered to have been made irrigable: 12,000. After a listing of the various costs, which seem quaint now, comes the *dramatis personae*, with the president in first place, of course, and then Ing. Adolfo Orive Alba, the head of SRH, followed by others. At the bottom is the name of the construction company.

On the large-scale air photographs and maps produced in the early 1970s, as a beginning on a new intervention, the previous canalization of the time of Alemán seems highly aggressive. Contour lines were drawn at one-metre intervals; they easily detail canal banks. Long, straight lines and star shapes at canal junctions slash across the webbed canalization of the Prehispanic wetland field complexes and allow only fragments of the original stream system to show through.

The actual hydrological impact of the SRH intervention was not very dramatic. The Basin still flooded, but the floods subsided earlier than they would have otherwise, sometimes leaving the fish and shrimp, which expected to enter a growth period in the rich stagnant waters, stranded in the grass. At the very outset the irrigation water was to have reached around the Basin to the eastern side, but the water never did get much more than about half way around the loop. Thus less than one-half of the 12,000 hectares "given" on the monument were actually being irrigated by the 1970s. This is when the SRH technical personnel address an hydraulic system that falls far short of what was once envisaged and a region deficient in many other ways as well. Rehabilitation must be planned for, and it is, but very little can be carried out. The power and funding of SRH are curtailed in the late 1970s, the name is changed to the Secretaría de Agricultura y Recursos Hidráulicos (SARH), and the will for further intervention is gone.

There was an ecological impact, commented on very sparingly and obliquely by the technical personnel but noticed clearly by the residents of the region. They reported to the author that flood levels tended to be lower and to drop more quickly; that the streams flowed faster and carried cleaner water. However, there were also fewer fish, and it was speculated sensibly by an elderly informant that their habitats had been disturbed, their nooks and crannies in the tortuous old river banks had been scoured out. In addition, contaminated irrigation water drained freely into the bottomlands. Open water never lasted very long, so that the ducks from Canada that had settled in times past now continued their flight. The effects of the intervention had been compounded by the demands of an increased population: people decimated fish stocks with the poison out of *barbasco* roots and with dynamite; they hunted out much of the game.

SRH Materials and Their Authors

There had to be huge caches of documents somewhere: after all this had been a governmental operation. An important early author on matters irrigational in Mexico comments on a "vicious bureaucratic practice": studies and data in the offices of the government's technical specialists are not compiled systematically, much less published (Herrera y Lasso 1919 [1994]: xl–xli). The situation was no longer quite so vicious, of course, but serendipity would be as important as tenacity.

From the outset of the investigation of prehistoric remains in the Basin, it was clear that in addition to our own oblique air photos we urgently needed good vertical air coverage. In a very speculative discussion at the Companía Mexicana de Aerofoto of a possible contract for such coverage, and topographic maps to go with it, the manager observed that they had already done all this for SRH. He showed me a roll of the topographic maps covering the Basin and surroundings at a scale of 1:5000, with contour lines at intervals of one metre! Copies of these maps and the photographs from which they were made became mainstays of all our various investigations of the Basin.

Sometime later I was able to locate the central repository of SRH studies of basins to be affected by hydraulic works. Study no.1 in that repository deals with the San Juan Basin. I had it in my hands and had agreed with the curator on a way of obtaining a copy the following day. Events intervened, including an earthquake; the repository was closed and moved. We have never been able to find that study again.

Our archaeology of bureaucracy eventually yielded a fairly coherent and certainly richly suggestive corpus. Magnificent additional pieces will probably come to light soon after this book is out, but there is enough to be getting on with. No site is ever completely excavated; no map is ever drawn at a scale of $1:1$.

The core of the assemblage is a series of three unpublished reports out of the 1970s dealing with a plan to rehabilitate the irrigation works of two contiguous irrigation districts, that of La Antigua and Actópan. These contain some 1300 pages of typescript and figures, from which we have abstracted what is relevant to the first of the two districts; it corresponds quite closely to the Basin. The earliest report is dated 1973; another, accompanied by a volume of appendices, appeared in 1976; a third report is undated but apparently was issued sometime around the middle of the decade.

There are no explicit relationships between the reports; there are a few reassuring references within them to the air photographs and topographic maps we found at the Compañía Mexicana de Aerofoto, but no evidence, unfortunately, that they were really used. The reports can be considered as largely complementary and contemporaneous; often they are redundant.

The authors are as anonymous as monks in a scriptorium. There is only one name, an administrator's, in all three reports, and he was almost certainly not an actual author, but I have cited him on the title page of this chapter. The work was clearly done by minions, passed on up the line, edited, with judgments added—repetitiously—and then proudly handed over at ceremonial occasions. The writers clearly had their various technical specialties. Such people usually had engineering degrees and are called *ingenieros.* The term is widely used as a respectful form of address for someone who is technically adept, whether an engineering degree has been earned or not. That is not usual in English; the meaning of *engineer* is parallel, but not the usage. We will hold ingenieros jointly responsible for this representation.

Field experience has left some images to go with the appellation. I have met high-minded and competent ingenieros. Those writing the reports allow sensitivity and compassion to show here and there. However, some I have known have been less well prepared, narrow in their purview, and peremptory with their infor-

mants. Among the campesinos, ingenieros are not always as chari-
tably described as addressed. Their field clothes are sometimes a
little too clean and stylish, they come and go in pick-ups, they do
not stay to listen, they dictate. When I first read the reports that
are basic to this analysis, I felt that I knew these men.

Supplemental Documents and Fieldwork

The reports of the 1970s, particularly those on which the initial intervention at mid-century was based, clearly had their antecedents.
We found one, a slim, concise, well-considered affair by an evidently
mature ingeniero and complete with a map of soil classes on a preintervention hydrographic base and two possible routes for the proposed irrigation canal (SRH 1947). There are a few additional items
out of various archives and a range of secondary materials. One
feels frustrated at not having been able to obtain more, but then
the authors of the main series of reports were not playing with full
decks either. We know of items they do not refer to and they cite
items we cannot find.

I was fortunate to have access to the results of ecologically conscious field observations on the eastern side of the Basin in 1989
and 1990 by a student of anthropology at the Universidad Iberoamericana in Mexico City, Ana Lid Del Angel-Pérez (1991, 1992;
personal communications 1989, 1991). The ingenieros say little
about the eastern side of the Basin, so her help is critical.

I spent time in the Basin myself in the 1980s, working on Prehispanic wetland agriculture; the prehistoric was often filtered by
the historic. There were exchanges, first by the way and then more
focused, with those involved in or affected by the twentieth-century
hydraulic intervention. This included people who had worked on
the excavation and maintenance of the canals, those who were using irrigation water, and cattlemen who benefited from the drainage
canals, as well as their workers and foremen. More recently, I have
discussed the intervention at length with a thoughtful civil engineer, Juan Pablo Martínez, who worked for SRH in this district for
about twenty years. And it was possible to spend several hours with
the octagenerian Ing. Adolfo Orive Alba, SRH's minister during the
presidency of Miguel Alemán.

Some earlier fieldwork is also relevant. In the early 1960s I
was preparing a dissertation on twentieth-century rural change in
Southern Veracruz. This introduced me to many aspects of rural
Mexico in general, the tropical lowlands in particular, and brought
me in close contact with various ingenieros. To many of my aca-

demic contemporaries and myself, as to the ingenieros, progress seemed possible and modernization, including drastic rationalization and mechanization, seemed desirable. Projects, especially integrated river-development projects, were promising—the issue was how to bring them off successfully. Much of this we have had to reconsider; we had our suspicions then. Politics were obviously subverting agrarian legislation and development. Spontaneous change seemed more effective than directed change in any case; the entrepreneur, on a private property or an *ejido*, seemed particularly admirable.

The General Conceptual Context of the Intervention

Behind the ingenieros' observations and assessments are certain imperatives. One of the strongest is the old aversion to wetlands and the conviction of the necessity of drainage, which is traceable at least to medieval Europe, articulated well regarding the lowlands around the Gulf by Humboldt, and expressed with eloquence by subsequent observers (Siemens 1990: xiv, 117–118, 120–121). The goodness of drainage suffuses the terminology of soil science (e.g., Wild 1988: 419–421, 835). It is basic to the ingenieros' classification and characterization of the soils of the Basin. A corollary is the desirability of irrigation, which also has very deep roots.

Various early-twentieth-century classics on irrigation in Mexico elaborate on the idea. Leopoldo Palacios, for example, points out that it is necessary to convince proprietors to substitute irrigated for rainfed agriculture, as well as to improve strains and increase productivity, so that the farmers can become more prosperous, raise the value of their own lands, and reduce the country's need to import basic products. Mexico will become a more important place in the world (1909 [1994]: 12). This is echoed in the reports of the 1970s and is not out of date yet, in the time of NAFTA.

Integrated river-basin development was the ingenieros' paradigm for the achievement of drainage and irrigation. Roberto Melville has provided us with a helpful guide to the application of this idea in Mexico (1994). For centuries the river basin had been taken most everywhere as a logical regional entity. The development of basins was immensely facilitated around the beginning of our century by technological development: the use of concrete, steam and electrical energy, massive dredges, and dynamite. And what was being undertaken in the Tennessee Valley between the wars became the great model of integrated river-basin development, including dams for flood control and the generation of electricity, river canalization, reforestation of surrounding slopes, agricultural development, promotion of cooperatives for various ends, and more. This vast project had a particular context in the United States that was not easily found elsewhere, but its promise became widely attractive and its approach was adopted in various countries, including Mexico.

The idea ignited politically in Mexico during the presidency of Miguel Alemán (1946–1952) and led to the formation of the Secretaría de Recursos Hidráulicos. A series of big basin projects were undertaken, each under the direction of a *comisión* within SRH, beginning with the Papaloapan, just to the south of our region.

The terms of reference of the intervention were succinctly set out by Ing. Adolfo Orive Alba (1947). He undertakes a comparison of patterns of food consumption and production in the United States and Mexico. It is obvious that the latter falls dismally short. Mexico has in fact been importing maize and other basic foodstuffs for many years, he points out; this must stop. The United States is worried about being able to supply its growing population; its experts predict a major food crisis by 1970 if massive new irrigation works are not undertaken. How much more, therefore, in Mexico! It is absolutely necessary to increase agricultural production in order to avoid a catastrophe. This can hardly be achieved by the various measures already being undertaken to improve agricultural production, which are all very important, nor indeed by further drainage in the lowland tropics, which is important too, but rather through a massive increase of areas under irrigation in many parts of the country, including the San Juan/La Antigua Basin. This was the gist of the rhetoric, and the large basin projects did draw the world's attention. The work undertaken in the San Juan Basin was actually a miniature of all that.

The reports on this project must also be seen against the schizophrenia of twentieth-century Mexican agrarian development, of which the graduate student wandering about in Veracruz in the 1960s was gradually becoming aware. Agrarian reform provided for group holdings, the *ejidos*, whose members, the *ejidatarios*, had only usufructuary rights to their parcels. Great emphasis was laid on the power of the ejidal assembly, but the government in fact assumed responsibility for guidance and assistance. A mentality of dependence had developed within the ejidos. But there was also a place in agrarian legislation for private properties, which were to be called "small properties" (*pequeñas propriedades*); there were to be no more large private properties—but of course there were. Entrepreneurship and commercial production were to be allowed to develop, under certain restrictions. On the one hand, then, there was great pressure to break up and distribute more large holdings, while on the other there was also a pressure to prevent further affectation and to facilitate commercial production. There were two main complexes of needs and possibilities, and it was often difficult to satisfy both at the same time. These two worlds emerge clearly from the reports and from ancillary materials; in spite of the recent changes in the agrarian code, they persist to the present time.

The ingenieros are determined to promote irrigated agriculture on private properties down to the smallest sizes and on ejidos. They do not concern themselves a great deal, on the face of it, with cattle ranching, especially on large holdings, although these and large sugar and mango plantations would be major beneficiaries of the hydraulic rehabilitation. On such holdings the favor of the wetlands and wetland margins had been realized for a long time. They are virtual historic constants in this landscape but do not have a commensurate place in the reports. It was obviously politically important to stress agricultural development on medium and small holdings and to remain discreet about the larger holdings.

Behind the authors of the reports are the "authors" of the Basin's cultural landscape, the politically powerful and wealthy people who obtained land for themselves in and around the Basin and who would benefit from improvements. One could write the history of the Basin—as of many, many other places in the world as well, of course—in terms of the interests and the estate building of a series of such men: Ruíz, with his two hundred slaves and series of *estancias* in the Basin; the Jesuit fathers who administered the huge Hacienda Acazónica; Santa Ana, the scoundrel *par excellence* of nineteenth-century Mexican history, who had his favorite hacienda, Manga de Clavo, not far from where Ruíz had his headquarters. By the early twentieth century certain family names had become prominent in the Basin. Soon several nationally prominent politicians obtained land in lowland Central Veracruz, none more prominent than Miguel Alemán himself, the "planificador y constructor de México" (Wise 1952: 56; figure 7.2). It is quite apparent, as Juan Pablo Martínez has pointed out, that the considerable sums of money that were spent on irrigation and drainage in the San Juan Basin would not have been spent if there had not been the potential of great benefit for prominent landholders (1994).

One can call this the *Sayula* factor, after the name of Alemán's Veracruzan estate. An imposing arch along the Veracruz-Jalapa highway, between Tierra Colorada and Puente Jula, carries the name; a fine approach, very much like the avenues introducing

Fig. 7.2. President Miguel Alemán (SRH 1976a: 159).

the large haciendas of the colonial period, leads to an airy villa on top of a fairly precipitous remnant of tertiary sediments. Its lands are choice—irrigable and not subject to inundation. The name is sentimental.

Miguel Alemán was born in Sayula, a small town on the isthmus of Tehuantepec, in 1903. He took a position in the federal Secretaría de Agricultura y Fomento in 1931, became governor of Veracruz in 1936, and then served as president of the country from 1946 to 1952. Between 1961 and 1983 he was the director general of the National Tourism Commission, a kind of ambassador at large. He died in 1983. His memoirs (1986) make interesting read-ing; they are mellow, well written, and include some good lines. His is a fine story and a rich life, a flood ridden on to fortune. The account humanizes the vast quantities of cement poured during his *sexenio* but helps us very little regarding the decision-making that shaped the intervention in the San Juan Basin.

Alemán's "authorship" emerges clearly from an admiring foreign assessment while he is still in office (Wise 1952). The powers given a Mexican president make him a king; this president is a strong king, a paragon: incredibly active, thoughtful, efficient, effective, un-tiring, able to command, vigorous—but virtuous! And he reveres his mother as a good Mexican man should. No doubt all this has been revised by someone somewhere, but the portrait is plausible for its time and the active mode it embodies informs our analysis. The president is planning and building when Wise's book is being written. Some of that momentum is still detectable in the SRH twenty years later when our reports are being written. Large, com-prehensive agricultural projects seem feasible and fundable. Tech-nically trained people still have answers. More cement must be poured. However, the momentum would subside and the terms of reference, even the name, of the SRH would be changed before the 1970s had ended. And it would turn out that the Mexican efforts in aid of irrigation would serve not so much to further the goal of self-sufficiency in food production, as to supply the U.S. market with fresh produce (Sanderson 1986).

The More Particular Predisposition of the Men in the Field

Most ingenieros will not have been from the area in which they were working. In the rural parts of the tropical lowlands this is general for professionals of all sorts; it follows from the way popu-lation and educational facilities were then, and still are, concen-trated in the central uplands. The prominent observations on the lowlands have long been made by outsiders, or *fuereños*. Most of-ten these are people who have come down from the highlands; the people of the lowlands sometimes call them, not without irony, *arribeños*. They bring a particular perspective, a view from above, and are generally judgmental and prescriptive; they also commonly have little empathy with the tropical lowland milieu.

Juan Pablo Martínez notes that the early field teams were made up of individual, mature engineers assisted by inexperienced assis-tants—*cadeneros*, or "chainmen." The 1947 SRH study has a ma-ture author, and there are patches of maturity in the later studies as well. With time and the development of training programs, which were an imperative of the Alemán presidency, the average age of the engineers came down. The reports of the 1970s were largely prepared, it seems, by such freshly baked ingenieros.

They were better qualified for some aspects of the kind of study required in the San Juan Basin than others. They had had little education in matters cultural, Juan Pablo Martínez assures us (1994); anthropology was not in their curriculum. They were thus neither equipped for, nor much inclined to, firsthand inquiry among local people. The SRH ingenieros' lack of consultation with local people on matters affecting them was to become notorious—and has apparently only just begun to change in recent times. But in places in their reports it is clear that they are not entirely insensitive, that they have been affected by the poverty they see around them in the Basin.

There are no Sartorian sketches here. The conventions of report-writing would not have allowed them, of course, but there is little evidence that the authors are holding back sensitive stuff. The reports were obviously finalized in central offices, and the ingenieros are aware of how this can wilt their observations; they protest repeatedly that they have been in the field, that they have observed directly. However, fieldwork by their own admission included a good deal of grubbing in the local offices of the project, for materials already compiled there. In their own direct field inquiry, they seem to have asked mainly presence-or-absence questions.

They include many photographs, but these do not help much. They are all ground photographs; it did not seem to occur to anyone to get above the obscuring vegetation, to do air reconnaissance and oblique photography. There is therefore little landscape; the most impressive images are near views of inappropriately irrigated fields or blighted crops.

The prose of the ingenieros has brighter and duller sections— long sections of the latter—according to the conventions of the various specialties represented and the style and inclinations of the individual. But there is one dominant and disturbing tendency: an extreme categorization of the material. Headings and subheadings abound; segments are numbered to the third and fourth order. There are many single-sentence paragraphs, abrupt shifts with scanty or no transitions at all. There are curious listings, percentages, and ratios. Vague phenomena are specified to the last digit and sometimes into decimals. One can understand that the ingenieros needed to generate certain specifics about the volume and kinds of earth that would need to be moved, the people that might benefit, what they would be able to grow, and especially how much more of it they could grow with irrigation. Such data were necessary in order that someone else up the line would be able to formulate an argument. But the attempts to be exact when the base could hardly be there become surreal.

We look for synthesis, perhaps even some ecology. The ingenieros are clearly hoping that if enough specific information is piled up, layer after layer—the geology of the Basin, the hydrology, the vegetation, then economic and social data, and so on—surely a coherent whole will result. Unfortunately, the material does not compost.

The Face Value of the Reports

This first pass will trace the explicit rendition of the Basin, which is virtually a landscape pathology. From prehistoric and historic evidence we have come to see productively linked microenvironments; the ingenieros see a place that is minimally useful for intensive agriculture and must be rehabilitated. Further, more ramified consideration will bend the reports and related materials to our own purposes, taking clues from them and adding field experience in order to appreciate more fully the Basin, its inhabitants, and its use in recent times.

Diagnosis

The negativity that permeates the reports begins in the assessment of the Basin's physical environment. The climate seems interesting mainly for its hazards; the soils are classified in terms of the impedimenta they present to irrigation. The very form of the Basin itself is faulty. It will not drain properly because of that difference between the level of the bottom of the San Juan and the La Antigua channels at their junction.

The regional water balance leaves more than enough water available for the irrigation of any land that can be made irrigable in the Basin—we are treated to many pages of tabulations and calculations to establish that. It is the installations; they are in a serious condition, here as in all the irrigation districts of the country (SRH 1973: 139)! The irrigation canals in the La Antigua and Actópan districts remain 80 percent unlined; some have been breached. Many sections are clogged with vegetation. Infiltration, transpiration, evaporation, and siltation have all had their effects. Water does not get all around the loop of the canal. In 1976, for example, it did not reach beyond kilometre 57.5 (SRH 1976a: 324), which is just about its southernmost point (figure 7.1) and which by then has served the substantial holdings on the west side of the Basin, including Sayula.

The irrigation water that is provided is not well used; the ingenieros find this galling. The physical facilities can be fixed; the correction of usage will be more difficult. The land on which irrigation is undertaken is often not sufficiently leveled, resulting in frequent overirrigation in the hollows and underirrigation on the humps, as well as excessive use of water in general as the users attempt to reach as much of their uneven terrain as possible. Fields are often plowed downslope and irrigated by means of those furrows, inducing erosion. The campesinos also do not know enough about the timing of their crops to irrigate optimally. And they will not irrigate at night to minimize evaporation. So, for one reason or another, only about 4500 of the 12,000 "irrigable" hectares claimed on the monument were actually irrigated in the years just before 1976 (SRH 1976a: 414).

Nevertheless, the ingenieros take courage from the fact that the possibility of irrigation has been well received, they are prepared to believe that the *espíritu temporalero*, the shifting cultivator's mentality, has been almost completely overcome (SRH n.d.: 5.1.2.); rural people are intensifying their operations and this is progress. However, the ingenieros note that much of the land that was initially cleared below the level of the trunk canal in expectation of irrigation water has been allowed to grow back into forest. Then there is also the forested terrain they note just above that, in hill lands to the west, and on the older, fixed dunes to the east. In spite of themselves they are seeing the concomitants of shifting cultivation— land necessarily left fallow with the moth-eaten look that indicates various stages of natural regeneration. So, one doubts that the spirit is yet quite so overcome.

Although the agriculturalists appreciate the irrigation facilities, such as they are, they are not happy, the ingenieros note, about the increased user fees that will come with rehabilitation, nor, perversely, will they be inclined to compensate the government for the improvements. But even if they were willing, one of the more perspicacious ingenieros observes, there is a fundamental economic problem. The La Antigua district, where irrigation is needed for a limited part of each year, cannot be self-sufficient; its income cannot cover the costs of the construction and maintenance of the works needed—a serious point, and perhaps basic to the lack of follow-up of all this diagnosis.

There are many more deficiencies in agricultural practice: The campesinos are generally casual in their use of fertilizers, other chemicals, and machinery. The image that inspires, in Mexico as in other developing countries of the world, is that of the man on a tractor pulling some implement across a vast level field (e.g., Maria Dorronsoro 1964: 102). Many of the men of the San Juan Basin still use oxen, which offends the ingenieros. There are so many cropping possibilities but a limited range prevails: grasses, sugarcane, and maize take up 82 percent of cultivated land in the La Antigua district (SRH 1976a: 415). The maize that is planted is of the inherited, genetically diverse kind. Little attention is being paid to more advanced plants and procedures.

The varied timing of maize crops is considered quite disorderly. This needs to be shaken down and scheduled uniformly so that machinery can be used efficiently between properties at the various stages of cultivation. This seems a gross misreading; we are conditioned to see subtle purpose in peasant practice. Staggered maize crops were one of the great strengths of old Mesoamerican subsistence systems.

Over and over again the ingenieros remark on disorder, as though they were nineteenth-century observers from abroad, criticizing the rural scene. "Traditional" practice is lost to them; they cannot take the usual greenery around the houses and along the roads and field borders seriously. There are many fruit trees but they are jumbled and not properly tended. Of the household animals, they note only the pigs, which they see as scruffy animals of very mixed breeds, but which they do recognize as ambulatory savings accounts.

Ranching, although sparsely treated, does not escape censure. It is seen as extensive; the breeding is uncontrolled, the animals are not separated properly by corrals when the time comes. Most ranchers still use bulls, only a very few have experimented with artificial insemination, and no one keeps a reproductive register. The mixture of centuries continues: *criollo* breeds predominate, as do *criollo* cultigens. It is all quite promiscuous.

Cattle are not stabled: the implicit models are North American beef-feeder and dairy operations. The cattle are not given silage, nor mineral or vitamin supplements; only a very small minority of ranchers feed molasses. There are good grasses in the Basin, but rotation of pastures is rare, as is the use of herbicides and fertilizers.

Overgrazing is the norm. The animals are sprayed against ticks, not dipped, which would be more efficient. Injections are commonly given against some diseases, but no one combats gastrointestinal parasites. Veterinarians are not usually called in; patent medicines are commonly applied—all very reprehensible, especially since there is a center of animal experimentation not far away, at La Posta, near Paso de Ovejas.

Rational ways of doing things have been developed during two decades of investigation at La Posta, as well as the agricultural research station of Campo Cotaxtla, to the south, and have been described in the technical literature. It is deeply frustrating to the ingenieros that modern methods are just not being used except on larger ranches and in the sugar industry. A woefully inadequate agricultural extension must be at fault.

Perhaps the problem is more fundamental: an insufficiency in the formation or education of rural people in these parts. They are not informed, they have not been "capacitated" (SRH n.d.: 5.1.2), they are not "cultured" in matters botanical and zoological (SRH 1976a: 415; SRH n.d.: 5.11). One reads profound impatience between the lines.

But, then again, someone says understandingly, there is not an adequate context for such "acculturation." The productive apparatus is deficient, there is much unemployment and subemployment, many are going to the cities or the United States to work. Education and other aspects of social life are stagnant.

There is little credit available; this was a refrain throughout my own fieldwork in Veracruz during the 1960s, as it was when the ingenieros were in the Basin. One of them gives the idea an additional twist. It is just not logical that there should be a lack of credit, that this should be such a brake on development, in an area where irrigation could assure reliable production year after year (SRH 1973: 140). The author is aiming at the policies of the banks, which had their own constraints. Their agents would explain to us that in order to be as accountable as politicians demanded of them, they could only lend against the titles of private properties and had to insist strongly on prompt repayment. The clients, particularly the campesinos, often saw this as tyranny.

Other things are unfortunate: insurance is rare; local roads are not what they might be; information on markets is inadequate or cannot be complied with. Instead, again and again, it is necessary to rely on intermediaries. They are the rural bogeymen. But the campesinos are also notoriously reluctant to organize, to form cooperatives, and thus perhaps to escape the stranglehold. Only the cane producers and the cattlemen of substance are organized. The campesinos never have any capital, they are always on a very narrow margin, etc., etc. I thought I was reading out of my own field notes.

No one moving through the Mexican tropical lowlands in the 1960s and 1970s could fail to notice unease over land tenure, erosion of the tenets of the agrarian reform, and general insecurity. The concentration of land in the hands of individual cattlemen is alluded to by the ingenieros; it is seen as an obstacle to agricultural development but is also tacitly taken as something insuperable, virtually a given. And rehabilitation will again help those most who are best capitalized. The fragmentation of private holdings into *minifundia* is noticed. However, invasion of private lands by ejidatarios, or the landless, is considered rare in lowland Central Veracruz; I found it common in Southern Veracruz at about that time. One of the great sins of post-Agrarian Reform Mexico, the renting of ejidal parcels to private proprietors, is not as serious in the Basin as elsewhere, either, the ingenieros tell us. In matters of tenure, then, things are quiet.

The reports give us many loaded words, flung as epithets: "traditional" above all others, "extensive" (which is just as bad), and "irrational," or even "vicious." There is much "disorder"; the plants and animals are mostly *criollo*, the grasses are only "natural." On every hand is *subaprovechamiento*, a failure to take full advantage.

Prescription

The remedies are already implied in the diagnosis and hardly need to be detailed, which does not deter the ingenieros from doing it, again and again. The recommendations are also vast, impractical, and, indeed, in some social and cultural respects preposterous— from our present perspective.

Most of the engineering recommendations are quite specific; they could have been carried out if there had been the money and the policy decisions: clean and line the irrigation canals, level the fields, build additional floodgates, extend the canals to serve the needs of landholders in neighboring *vegas* (the river bottomlands to the west), put in webs of canals or tubing to drain individual parcels effectively, straighten and deepen the streams further so that they drain more rapidly—to the extent that the fundamental hydrological problem of the level of the Basin's exit allows.

SRH envisages integrated basin development; a whole range of expedients is thus prescribed, from the correction of the physical environment to the capacitation of the inhabitants. Many specific recommendations are made regarding agricultural and ranching practice, which are already much more complex and difficult than the pouring of cement: crops must be diversified, adding soy beans and sorghum; nitrogen-fixing crops must be interdigitated with highly demanding crops such as sugarcane and maize; in an area ecologically suited for the production of fruit, orchards must be developed; on the ranches, natural grasses must be replaced with improved grasses, the animals stabled, the breeding controlled.

Among these recommendations there is one that is quite interesting. Cropping and animal husbandry should be integrated. It is likely that the ingenieros were reading some of the literature on the advisability of transferring old European peasant wisdom regarding manure into the tropics (e.g., Waibel 1950; Gourou 1966). They seem to have missed the integrated crop and livestock management already developing on ejidal and small private holdings, as will be outlined.

The ingenieros see the necessity of institutional changes, of improvements in the economic and social context of the inhabitants. With this, their prescriptions become vaguer and vaguer, eventually washing up on one imperative: It is necessary to improve the culture of the people of this Basin (e.g., SRH 1976a: 264).

In all of this science remains a fundamental enthusiasm—epitomized, perhaps, in strong suggestions that hybrid maize should be adopted. Alemán himself is much taken with this great new thing, the advantages of which he sees as plainly evident. He is photographed, smiling, with a cob of it in his hand (Alemán 1986: 284, 286).

We have learned to hesitate on this and other related issues. Alemán and his ingenieros seem naive and overready at this remove. They swear up the whole Green Revolution with gusto; we have long since seen the unfortunate results that it can have. We have come to think of lowland tropical nature, as well as human adaptation to it, as having much to teach us, as not so disused and not so backward as the ingenieros affirm.

Little came of all their recommendations, directly. The reports, the air photographs, and the magnificently detailed maps seem, in themselves, the most important residue. Change occurred, of course, but at its own pace and hardly as a result of this planning. Aside from that, residents of the Basin have recently been encouraged by government functionaries to take responsibility for the planning, execution, and financing of change.

A Closer Analysis of the Ingenieros' Observations

The pathology represents the Basin in the ingenieros' terms, but their reports and related materials suggest more. They "produce" a clear region; it is too clear. They treat its inhabitants shabbily; that needs to be reflected on. They are quite eloquent on the "land" once some of their materials are distilled, in that they give us a further reading on the geography of favor. If we ask again about discontinuities and continuities in land use, we hear less of the former and more of the latter than might have been expected.

Regionalization of the Basin

The key expression of both the ingenieros' achievements and their representation of the Basin is in the great crenelated loop of the main irrigation canal (figure 7.1). It is a clearer regional boundary for the San Juan Basin than we have had in any of the earlier historical materials. The area thus enclosed is the early irrigation district of La Antigua, comprising the irrigable land south of the river, which is sometimes considered separately in the ingenieros reports but often together with the Actópan district just to the north. Rehabilitation was to be attempted in the two at the same time. Later the two would be lumped together under the name of La Antigua and administered out of Cardel, the main commercial and service center of these parts. In any case, this clear, tempting boundary is often used too categorically. Essential local relationships to neighboring areas upslope are thereby neglected.

The grand altitudinal scheme of environment and human adaptation is not there. The ingenieros have either never been made aware of it or just do not consider it relevant. They are concerned with relief below the line and at a very large scale, mostly as something to be leveled in order to facilitate gravity irrigation.

Regionalization in the reports goes to another extreme when the ingenieros resort to data by *municipio*, the official Mexican census statistical unit (figure 7.3). This is a basic challenge to the social sciences in Mexico, certainly in any inquiry that includes ecological elements. The municipios vary tremendously in size and configuration. Political considerations dominated in their delineation. The boundaries cut cavalierly across physical environ-

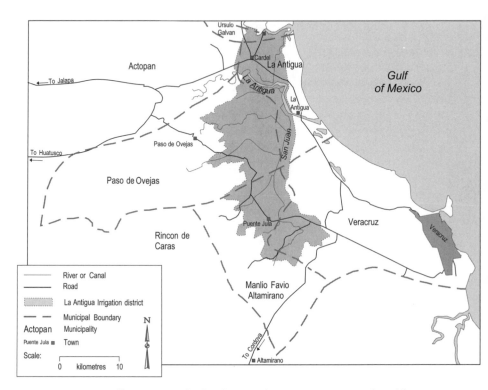

Fig. 7.3. An illustration of a fundamental Mexican statistical problem: a superimposition of the *municipios* of Central Veracruz on the entire La Antigua irrigation district (a compilation).

mental variations and, indeed, most logical human geographical regionalizations, too.

Representation of the Lowlanders and the Way They Live

The lowlanders are vivid in earlier sources: Indian woodcutters in the forests out on the periphery of the world as seen from Old Veracruz; their *caciques*, who can be easily brought around with a jug of wine; Spanish priests, traders, and *hacendados*; black slave stevedores and carters. Sartorius sketches us a colorful lowland subset of the full Mexican array, lingering on the ranchero, the vaquero, and the amazing *capataz*. The lowlanders in the ingenieros' reports are cardboard cutouts.

We have deduced that there are at least 25,000 people in the settlements in and immediately around the La Antigua irrigation district in 1970 (figure 7.4). There are no comparable regional

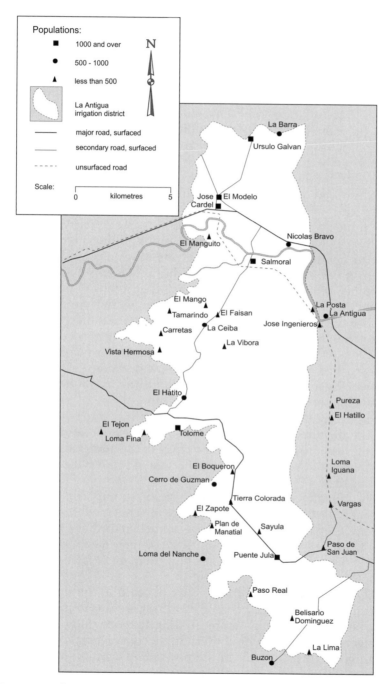

Fig. 7.4. Settlements in the San Juan Basin in 1970 (a compilation).

figures out of the colonial period or the nineteenth century to be set over against this. In the *municipios* that encompass the district, population is increasing at about 2.8 percent per annum, which is about the state average (SRH n.d.: 4.14).

Most people in the Basin of the 1970s, we are told, are in primary production. However, many rural people in the Basin are "subemployed" or "unemployed." There are few opportunities to earn additional money. And when something becomes available, like clearing the woody growth out of pastures, day laborers get less than minimum wage. And this day labor, the ingenieros observe, is still reckoned not by the hour, which would be the modern way, but by *tarea*, by task, and this often requires ten to eleven hours a day to complete.

A minority work in some aspect of the sugar industry. Sugar harvesting yielded some strong impressions during my own Veracruzan fieldwork: loaded tractor trains moving toward the mills, the massive scale of the crushers at the mills themselves, the smell of molasses all around, but more than that, the very hard work of cane-cutting, the columns of workers coming in from the fields at the end of the day, black from the singed cane that had been burned before harvest, coming home to lodgings that reminded me of pictures out of Nazi concentration camps. The rebels out of the Sierra Maestra had just "liberated" Cuba; we were reading about the tyranny of sugar, as a crop, especially the "dead time" it imposed on a landscape between harvests. There is very little of all this in the reports, except that sugar is seasonal; sometimes other cropping can be interdigitated with it, sometimes not. Sugar harvests begin late in one year and go on into April or May of the next. Rain-fed maize can be planted just after that and harvested before cane-cutting begins again.

The ingenieros also note that all of what rural people earn goes for basic necessities, most of it for food. This food, as well as the clothes they notice and the shoes, are deficient. They are affected by this; it enters their constrained prose repeatedly, perhaps indicating that for many of them the improvement of facilities for irrigation was a mission.

They present a table of the causes of death in the state of Veracruz in the 1960s by descending percentages—unadorned with any discussion, but perhaps more striking for that (SRH n.d.: 4.14). Someone cited the causes of death in the four hospitals of the town of Veracruz in the middle of the nineteenth century (Mayer 1844: 7, cited in Siemens 1990: 93). At the top of the list in both cases are diseases of childhood and the digestive tract. These are the diseases of poverty, of poor hygiene and poor housing. But, our authors of the 1970s note, public health is improving, little by little; one town after another is getting good drinking water. Also, death rates are falling while birth rates remain high: in fact, a "demographic explosion" is underway (SRH 1973: 21).

Little attempt is made to characterize the population. The ingenieros do not dare or care to say anything about the race of the people of the San Juan Basin. Central Veracruz is probably just taken as obviously a region of rich racial mixture. They do consider what brings people together in this region. This was always an important element in the thought of the 1960s and 1970s about development and colonization in newly opened lands: social coherence in the face of a range of practical problems. In this more than anything else, the campesino seemed obdurate. Our authors note, disapprovingly, that the population is very individualistic, suspicious of organizations, and, in any case, preoccupied with practical concerns; only 7.1 percent of the population of the project are affiliated with political parties. Only 26.7 percent of the agriculturalists are participants in producers associations, and only 12.8 percent of those who have cattle belong to cattlemen's associations. Within the latter, however, there is a great difference between the private proprietors and ejidatarios—45.1 percent of the first and 6.2 percent of the second belong.

We are given a few paragraphs and a table on *practicas tradicionales*, by which is meant, mostly, religion (SRH 1973: 25–26). The categorization and the specification into at least the first decimal point continues. They have found by means of their field inquiry that in the area of the project 37.4 percent of the families participate in religious ceremonies. Rural participation is considerably higher than urban, and so on. Here and there out of the column headings one may deduce some prayer and dancing, but mostly the ingenieros' representation passes over it all statistically and aridly. There is no jarocho anywhere.

Assessment of the Land

The main objective here is to tease out one more situated assessment of the geography of favor in the Basin, and that from material that in its explicit, highly categorized form hardly coheres. This geography is to be set alongside those deduced from prehistoric remains and the more or less specific historical materials. We will look first at how the land is held and then what it is like (table 7.1).

Table 7.1. *Land Tenure (Source: SRH 1976a: 15–17).*

A. Holdings in the La Antigua District:

	Hectares	
Private properties:	10,151	(349 *productores*)
ejidos:	9,827	(39 *ejidos*; 1746 *ejidatarios*)
Zonas urbanas:	458	
Total	20,436	

B. Sizes of private properties, numbers, and areas of such holdings in each size category:

	No. of properties in each	*Total ha. in each*
0–20 ha.	308	2,187
20.1–50	93	2,943
50.1–100	46	3,153
100.1–200	9	1,094
200.1+	2	774
Total	458	10,151

When the areas occupied by the settlements are set aside, the two rural worlds emerge: the private and the ejidal. The size of the individual ejidal parcels in the district, a very important figure anywhere in rural Mexico, is on average 5.6 hectares; that is, the 9,827 hectares of ejidal land divided by the 1,746 ejidatarios, which is not bad as far as ejidal parcels go. In more densely settled areas in the lowlands or in the uplands the figure would be lower. Most of the private holdings are on the low end of the size scale. It has been maintained that in Central Veracruz, generally, uneconomic fragmentation of holdings (*minifundismo*) is a more serious problem in the private than the ejidal sector (Cisneros 1993: 71). Many private proprietors have more than one property (458 holdings per 349 producers), more probably on the lower end of the scale rather than the higher. The numbers of property holders go down as the property size goes up, of course—that is a staple of the human condition. The reports handle these holdings on the upper end with care, understandably. No names are given, of course, but it is observed that both of the biggest holdings are over the limit that the Agrarian Reform Law allows (SRH n.d.: 4.51). And one author cannot resist mentioning that among the private proprietors there are some *exfuncionarios públicos* in whose hands are concentrated large parts of the best land (SRH 1973: 71–72). From what we know of the lowlands in the 1960s, it is likely that such holdings are disguised by multiple familial titles or are covered by certificates of inaffectability, which had been handed out by presidential decree to many large landholders in the preceding decades.

The Agrarian Code, before its reform in 1992, hardly recognized large private properties; private properties were mostly referred to as small properties (*pequeñas propriedades*), regardless of size. The code also forbade the renting out of ejidal lands. They were not to gravitate again into private hands. The emphasis given this problem varies within the reports, no doubt according to the ideological inclinations of the individual authors, but it is apparent that it does take place on a considerable scale in the 1970s. Better irrigation facilities, one ingeniero notes, should make ejidatarios, or for that matter private proprietors who have rented out their land, want to work their own parcels again (SRH 1976a: 18).

The ingenieros make much of the soils of the Basin, as might be expected, but it is not easy to use their pedological information here. Given their concern with the specifics of irrigation, it is logical that the ingenieros should establish, map, and describe soil series, the most detailed level of the currently predominant soil classifications (Wild 1988: 815–843). The maps are very detailed and, in the form in which they reached me, also largely indecipherable, but the descriptions are useful. There is a curious, helpful profile that links topography and the soil series (figure 7.5).

The ingenieros also need a pragmatic reading of soils, particularly with respect to slope, texture, and drainage. It makes sense, therefore, that they should survey and map "land capability" (Wild 1988: 831–832); i.e., organization into classes on the basis of limitations *vis-a-vis* a particular objective, a desired land use. We have a classification along these lines for 1976, but no map. Fortunately there is a somewhat simpler land capability classification for the Basin, prepared in 1947, that is accompanied by a map, but it is within a preintervention hydrological context.

There is no practical way to reconcile these two attempts cartographically here. The expedient will be to reduce considerations of land capability into simple categories, link these to soil series designations as on figure 7.5, and provide descriptive locational references that draw on the foregoing.

The Best Agricultural Land.   Since early in the history of agriculture, low, level, naturally humid and fertile land on the margins of

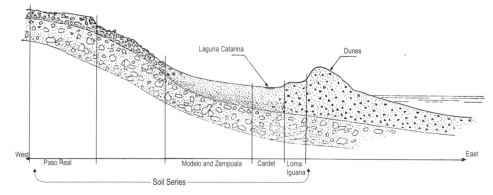

Fig. 7.5. A schematic profile of topography and geology, with soil series referred to in the text (SRH 1976a: 118).

rivers has been considered prime wherever it occurs. In the Mexican context this is *tierra de humedad*, that is, land with residual moisture in the dry season; the riverside plains on which it occurs are commonly called *vegas*.

In the 1947 and 1976 classifications of the Basin soils according to their capability for irrigation, *vegas* soils are considered without impediment: first class (SRH 1947: 10). They are classified in other, more strictly pedological terms, as well, of course. On a recent map of soils in Central Veracruz, they are called *fluvisoles*, that is soils with little evidence of horizons in their stratigraphy, formed recently on sediments carried in by the major streams and laid over beach sands (Cisneros et al. 1993: 26–28; Hebda, Robertson, and Siemens 1991). The ingenieros fit them mostly into what happens to have been named the Modelo series.

Vegas/Modelo soils are dark gray-brown in color, sandy clay in their composition, and highly porous, i.e., easily permeable. They show only a minimal differentiation of horizons; the A horizon may be almost half a metre thick. There is some compaction in places but no impermeable layer. These are relatively "azonal" soils, to use an old designation: the parent material is too fresh to allow for the development of a full stratigraphy (Wild 1988: 820). Their topography is not quite flat either, but rather gently inclined. The schematic profile (figure 7.5) shows them inclined toward a river; they can also slope gently away from it in the typical profile of a levee. Vegas/Modelo soils are humid by reason of seepage from neighboring streams, or unlined irrigation canals. The other side of the same coin is that they are well drained. They can be relatively easily irrigated to supplement the residual moisture in the dry months, and in fact are already mostly under irrigation in the Basin. Normally they flood only in times of very high water. We can place them in the lower reaches of zone *a* on our own model of wetland zonation (figure 1.9).

When Laguna Catarina reasserted itself strongly in 1993 (figure 2.6), we observed that the floodwater stayed just clear of all the main areas of Vegas/Modelo land. At the same time, most of the remains of Prehispanic planting platforms and canals were flooded. The water was tracing out a critical boundary of prehistoric and historic times, that between zone *a* and zone *b* on our schematic profile (figure 1.9).

Five concentrations of first-class soils appear on the western side of the Basin as mapped by the SRH in 1947: on the alluvium of (1) the Paso de Ovejas and La Antigua Rivers, (2) the Arroyo Tolomé, (3) an unnamed arroyo east of Cerro Guzmán, (4) two other tributaries of the San Juan, the Zopilote and the Puente Jula, which drain the Sayula lands, and (5) the upper San Juan itself.

The first three of these concentrations are mostly carpeted with cane, grown on small holdings under contract to a mill. That most of these holdings are ejidal is surprising at first, since it is a cliché of the history of Mexican agrarian reform that the best lands remained in private hands. How did this happen? It might bear investigation. It was here, on these soils, that the progressive hacienda El Faisán was located. Its cattle did not suffer the dry-season hardships of cattle on other neighboring haciendas with slightly higher land and nonalluvial soils. On El Faisán the pastures were always green (Skerritt Gardner 1989b: 124–125).

During various overflights and brief field inquiries in the ejidal communities on the western side of the Basin during the 1980s, it became apparent that agriculture there, on and between the hilly outliers of Tertiary sediments, is areally and seasonally complex. Multiple cropping is possible in many places on *tierra de humedad.*

Sayula, located on such land, is famous for its vast mango orchard, visible from the main highway. The ingenieros note other commercial cultivation of fruits and vegetables on prime land. From their vertical air photographs one can deduce ranching, too, at least for the southernmost of the concentrations. Most of this land is already irrigated, but the ingenieros see definite possibilities for making this universal, for intensifying the use of this land further. And they add that lateral canals could be extended from the right side of the main irrigation canal to serve *vegas* that extend into *barrancas* to the west.

They also notice an aberration: a small percentage of the Vegas/

Modelo soils, something like 1 percent, is impeded in its drainage. There are small hollows, in places, where water brought by local run-off, seepage, or irrigation gathers and remains for a time. It is stored there, one ingeniero says, without realizing quite what he is saying (SRH 1976a: 199). On the air photographs and large-scale topographic map there are several interesting swarms of these depressions, most with neighboring mounds (figure 2.20). These are large and little-studied prehistoric settlement sites. One can easily hypothesize that the depressions were excavated for building materials and lined with clay for water storage, as in Mayan water collectors, or *aguadas*. They will have been very useful sources of open water for a good deal of the dry season, and at the very nadir of the dry season, when the water level dips below the surface, they may even have been used for gardens.

In those times, as often happens now around growing population concentrations fronting on alluvial land, people seem to have built on their best lands, which we have come to see as ominous. One could give it a more positive interpretation. These are excellent sites for habitations that are only loosely agglomerated, where there is land in the interstices and all around for intensive multiple cropping—the lower edges of zone *a* (figure 1.9)—and tremendous possibilities in the neighboring zone *b* during the dry season.

There are other pedological variations on alluvial lands, particularly on the very margins of the through-flowing rivers, the La Antigua and the Paso de Ovejas, both within the Basin and in the *barrancas* upslope as well. The ingenieros discuss them as the Zempoala series; they take up just a few percentage points of Basin lands, but they are intriguing. They are found mostly on riverine terraces of various levels below the tops of the levees. At one point along the Paso de Ovejas one can count four or five such terraces in the alluvium within several hundred metres of the river's channel, each fringed with the cane grass that the vicar found impressive. They are subject to various degrees of inundation: the highest ones only during occasional extraordinary highs in river levels, the lowest ones every year. Such soils are crude, i.e., sandy, with stones and gravel; they are immature, with only a slight A horizon, and are well drained but vulnerable to erosion.

Some *tierra de humedad* is so often and so long flooded that it is left in scrub forest; some of it can be exploited in the depths of the dry season only. In the very lowest alluvium in the bottomlands of the Basin and in the neighboring *barrancas* it is possible to practice *agricultura aventurera*, a matter of slipping onto briefly exposed alluvium with a rapidly maturing crop, a risky kind of agriculture that is practiced according to need and opportunity, that may yield and may not.

Lesser Agricultural Land.   On the eastern side of the Basin there is a band of terrain that slopes up gently from the bottomlands into low hills. The soils on the lower reaches of this continuum are named the Loma Iguana series. They have developed on marine sediments that make up the older, lower dunes. These are fixed by low and largely leguminous vegetation. This terrain must be distinguished from the more recent, higher and still-moving dunes, which are sometimes used now as sets for films that need a desert backdrop. The lowest of the Loma Iguana soils may be inundated when Laguna Catarina strongly reasserts itself.

They are deep soils, with well-developed A horizons and B horizons that may go down to two metres. They are made up of sandy clays and may show considerable compaction, but not cementation. They are easily permeable; indeed, drainage presents no problem—which is always given importance in the ingenieros' analyses. But they are somewhat too well drained, not as fertile as the Modelo soils, and subject to erosion. They are also attractive to ants. Vast areas are pocked with the round cleared spaces, several metres in diameter, that surround the openings to subsurface colonies, which show up dramatically on air photos as swarms of white dots. They take up much cultivable space, and the ants themselves threaten crops.

Loma Iguana soils are also diminished in their usefulness by limited possibilities for irrigation. The great looping irrigation canal brought water into the southern extreme of this region briefly after it was newly constructed. There was a good deal of clearing in anticipation of further possibilities, but then the system failed. Spray irrigation, using water from wells and the river, has been installed in some areas. The ingenieros note that leveling would be useful in many places and not too difficult, but delivery of water is the problem. The land that is served by irrigation may be used for cane or for pasture. The rest is used mostly for rainfed maize cropping or seasonal pasture. Air photographs show extensive parcelization, and field inquiry indicates that this is largely ejidal land.

Soil of the Paso Real series covers much of the hilly terrain interfingered with *vegas* on the southwestern side of the Basin, and also the "islands" in the bottomland, just downslope. This soil has developed from weathered volcanic materials and hence may be rocky, as the ingenieros' profile indicates (figure 7.5). We found basaltic bombs around a milking shed on one of the "islands." Paso Real is a sandy, clayey soil; its horizons are fairly well developed and there is some compaction as well as cementation. The soil thus presents considerable resistance to plant roots but is not completely impermeable; however, the cemented strata are closely fissured.

Much of the land in this area was also cleared in preparation for irrigation. From what is said of the remaining natural vegetation, one can deduce remnants of the savanna that Sartorius describes. The considerable relief makes leveling necessary for gravity irrigation; some ranches have already installed spray systems. Almost any crops can be grown, but erosion is a potential problem. As it is, most of this land is used for cattle. This is wet-season pasture and part of the local, rotational management of grassland: the animals are moved downslope in the dry season, wherever possible, and back up after the rains. In this context dairying is integral to ranching with dual-purpose herds. Corrals, drinking troughs, and milking sheds are typically concentrated on topographic eminences safely above flood levels.

These eminences have been key points in the Basin and its surroundings for a very long time. They were habitational sites in prehistory, loci for *estancias* in the sixteenth century, and pastures for the individual herds, with their particular vaqueros, in Sartorius's time. From the air now one sees many trails converging on their whitewashed facilities. These are the "central places" of ranching.

Bottomlands.   Although the ingenieros do not refer to altitudinal zonation explicitly anywhere, they do work with it implicitly at a large scale, and sooner or later in their treatment of the various soils, there is some reference to a relationship with the bottomlands—the "poorly drained" soils just slightly downslope from the *vegas* or the hill lands on either side.

The soils of the bottomlands are considered the most impeded, the least tractable, and thus in the lowest categories of the land capability classification. There is the distant possibility of intensification, of course; drainage and other treatment could move these soils up the scale. As it is they are fit only for dry-season pasture. We have tended to fold back the ingenieros' categories, impressed not only with the patterning in the upper margins of the bottomlands and what it means with respect to ancient intensification, but also the way the cattlemen from early colonial times to the present have ranked the bottomlands generally. For them they are prime: they provide dry-season pasture, which is lush pasture where animals are fattened for sale. No ranch in this region could do very well without access to such land. The Basin thus has various sorts of "best lands."

The soils of the bottomlands are described as the Cardel series. They developed on marine clays enclosed by dunes and bars and topped by alluvium. The soil is sandy clay in varying proportions. The A and B horizons are fairly well developed, and the ingenieros see some subdivisions in each. The first may be as deep as thirty centimeters and the second, two metres. Clods develop on the surface; there is a good deal of compaction farther down, but no cementation. This soil has many cracks during the dry season, but these close up rapidly with the coming of the first rain. Soil scientists have a marvelous name for this process: argilopedoturbation. In the FAO soil classification, the resulting soils are called Vertisols (Cisneros et al. 1993: 26–28).

The distinctive characteristics of this soil, according to the ingenieros, are its dark gray to black B horizon, with a gritty to crumbly clayey texture, and low in that horizon a prismatic or even columnar structure, commonly called *cascajillo*. This soil drains slowly and only seasonally by reason of its topographic placement; one could also say that it stores water very effectively.

The bottomlands receive run-off from irrigation on adjacent land during the dry season, carrying salts with it. These go below the water table and can come up to the surface through capillary action, creating a problem of salinity.

The ingenieros get our full attention when they note that in one of the representative test pits of Cardel soils, between sixty and one hundred centimeters below the surface, ceramics intrude—no more, no indication of what kind, no speculation on what that might mean, just a notation in a column in a table. Even so, it is a corroboration of our own finds in similar locations at roughly the same levels (Siemens et al. 1988).

In our test pits we would see mostly subtle and frustrating gradations of soil tones within a fairly narrow tonal range—shades of gray, more or less, with brown admixtures. We strained our vocabulary to describe the shades, we looked for clear signs of construction, but we reluctantly realized that there was little definite visible stratigraphy here, natural or anthropogenic. Instead we had to rely on physical and chemical analyses.

The clay that dominates these soils lends them plasticity and adhesiveness, the ingenieros say: indeed it does. It cannot be dry-sieved; we took to kneading it through our hands in order to isolate the bits of ceramics. The clay sucks the boots from one's feet and sticks fast to spades and sampling knives.

Agricultural Use and Misuse

Table 7.2 shows the ingenieros' measure of the misuse of land and water in this Basin. Out of its 20,000 hectares, 18,000 are "dominated" in the 1970s by irrigation facilities (in 1950 it was 12,000) but only about 5,000 hectares are actually irrigated; 3,000 of that in crops and the rest in pasture. Rainfed agriculture is noted on something over 1,000 hectares, to which one should add what may well be fallow in the forested areas and steep slopes—perhaps another 1,000 hectares. The *espíritu temporalero,* the tendency or

Table 7.2. *Land Use in the La Antigua Irrigation District Early in the 1970s (Source: SRH n.d.: 1.74).*

Use	Hectares
Rainfed agriculture (on land that is within the area "dominated" [topographically] by irrigation facilities but where they are not actually functioning)	1,146
Irrigated agriculture (full extent ever irrigated, including actual irrigation at time of survey [ca. 5000 ha.], which, in turn, includes irrigated pasture [ca. 2000 ha.])	9,146
Unirrigated pasture	4,644
Forest (various heights and compositions)	1,621
Orchards	1,258
	17,815
Townsites	649
Riverbeds	368
Steep slopes	1,321
Total	20,153

need to rely on shifting cultivation that the ingenieros regard as retrograde, is not dead. Irrigated agriculture, on the other hand, is far below what it might be. Ranching is the main thing: one-third of the total area (2,000 irrigated, 4,600 unirrigated, plus some of the forest and steep slopes). Only one quarter is used for crops (3,000 irrigated and 2,000 otherwise).

From the vicar and the médico we get a still life of Basin crops, from Sartorius a cornucopia. From the ingenieros we get only a grudging table and commentary (table 7.3), but the productivity is still apparent.

One wanders through the numbers imagining failures in the irrigation system, disputes, market variations, droughts and floods, good years and bad; surely the curiosity of the ingenieros will have stretched as far as to induce a discussion of some of the more dramatic yearly variations, but it has not. The aberrant figure for grasses in 1973–1974 is probably a typographical error.

The ingenieros do allow themselves some speculation on new cropping possibilities. Sugarcane is the main crop, with more promise still for the future. It has already been a means of introducing advanced technology. It requires modern inputs, which in turn require the acceptance of technical advice. Tuberous crops, especially *camote* (*Convolvulus batata*), are grown in home gardens, do well in the dry season on well-drained soils, and should be expanded. The experimental station at Cotaxtla has new varieties available. The ingenieros argue for chile and for the tomato, other good dry-season crops, for multiple-cropping of beans, for a type of sorghum that is used for the production of brooms, and for soya, of course, the great crop of the future. Some fruit trees are already widely grown in respectable plantations, particularly mango and papaya. They are, however, highly vulnerable to *nortes*. More windbreaks are needed! Then there are all those other fruit trees jumbled up around the houses and along the field boundaries that must someday be properly commercialized. There is acknowledgment of the banana, that old scandal, which does well, as it always has.

Someone in the corps must have had entomological inclinations: he is given his chance and we are given a burst of Latin (SRH n.d.: 26.36). The lowlands are a difficult environment and there is a threat to every leaf and stalk. Someone else is conversant with weeds, and we are given another burst (SRH 1976a: 2.36). The seeds are spread easily by irrigation water. But there is hope in both respects, since many agriculturalists are learning to use chemicals— lots of insecticides, not too many herbicides or fungicides yet. And

Table 7.3. Crops in the La Antigua Irrigation District, in Hectares (Source: SRH n.d.: 2.9).

Crop	Year 1969–1970	1970–1971	1971–1972	1972–1973	1973–1974*
Avocado	7	5	6	6	—
Sesame	2	—	—	—	—
Pumpkins	1	3	8	7	—
Sugar cane	1,093	1,054	1,245	1,645	1,820.03
Green chile	59	49	96	43	43.64
Beans	182	172	78	46	178.04
Cabbage	—	1	—	—	—
Jitomate	—	—	51	53	32.00
Maize	884	945	765	788	874.71
Mango	66	55	59	69	—
Papaya	125	154	75	43	332.66
Grasses	1,962	1,915	1,807	1,786	131.00
Coco palm	1	1	—	—	—
Cucumber	25	69	45	58	72.77
Watermelon	7	—	11	—	—
Brown sorghum	14	12	4	5	3.60
Forage sorghum	—	—	12	13	—
Tomato	93	53	—	—	—
Carrot	—	—	1	—	—
Vegetables	—	—	—	—	6.89
Fruit trees	—	—	—	—	0.50
Citrus	259	221	171	247	35.00
Lettuce	—	—	1	—	—
Total	4,780	4,709	4,435	4,809	3,530.84

* Preliminary

regarding fertilizers, thanks to the advantage of proximity to an agricultural experimental station and suppliers in nearby towns and in Veracruz itself, they are being widely used.

Seasonality is not very satisfactorily treated in the reports, the metronome is muted, but there is a calendar of cultivated crops (table 7.4).

There is a substantial potential for multiple cropping, we are told, but the campesinos do dry-season cropping on only 10 percent of irrigated land; they could do so much more. As to maize, the ingenieros give the advice that the times of seeding and harvesting

Table 7.4. *Approximate Calendar of Annual Crops (Source: SRH 1976a: 373–374).*

Crops	J	F	M	A	M	J	J	A	S	O	N	D
Rice	—	—	—	—	—	—						
Rice							—	—	—	—	—	—
Sesame							—	—	—	—	—	—
Peanuts					—	—	—	—	—	—	—	—
Squash	—	—	—	—	—					—	—	—
Onions	—	—	—	—	—					—	—	—
Chile Jalapeño	—	—	—	—	—	—				—	—	—
Chile Serrano	—	—	—	—	—	—	—	—		—	—	—
Beans	—	—	—	—	—							
Beans							—	—	—	—	—	—
Tomatoes	—	—	—	—	—	—					—	—
Maize	—	—	—	—	—	—						—
Maize						—	—	—	—	—	—	
Melon	—	—	—	—	—							—
Cucumber	—	—	—									—
Watermelon	—	—	—	—								—
Sorghum	—	—	—	—	—	—						
Sorghum	—	—	—	—	—	—	—	—				—
Soybeans	—	—	—	—	—	—						
Soybeans	—	—	—	—	—	—						—

should be synchronized according to optimal timing worked out long since at the experimental station, in order to make better use of machinery. We have been conditioned to regard variation in the genetic make-up of common maize, the range of possibilities in the timing of the use of different varieties, as an important aspect of the diversity inherent in "traditional" food production, as a strength and not a weakness.

Basal Agriculture

In the very targeting of retrograde practices one detects the undertone of subsistence production that was strong in Sartorius's discussion of agriculture. Skerritt Gardner hears it in his late-nineteenth- and early-twentieth-century sources for the northwestern quadrant of the Basin; he notes a virtual definition of subsistence: the production of household necessities plus a marketable surplus (Skerritt Gardner 1989b: 77, 88, 90–91). Del Angel-Pérez analyzes what remains of it in the surroundings of Vargas.

The undertone is muted in the reports not just because of the ingenieros' lack of sensitivity to it, but also because the actual fund of traditions has been diminished, what with the discontinuity consequent on in-migration and modernization. The ingenieros note that there is still the "disorder" of household gardens, but from them and recent field observation it is clear that the objects of many other traditional subsistence activities—the fish, game, and useful wild vegetation in the surroundings—have been reduced through overexploitation, clearing, and drainage.

The undertone is detectable in some of the ingenieros' overall figures on agricultural production: 21 percent is for *autoconsumo* and 79 percent for the market (SRH n.d.: 5.12–13). Further integration into the market is clearly implied to be desirable; the relatively good roads of Central Veracruz help in this regard, as do the intermediaries, the buyers with their trucks. It is true that this brings speculation, but it also animates the producer. Interesting to see that the ingenieros can put a positive spin on that old impediment.

Production for *autoconsumo*, as though it were a debility, is most notable among those with very small properties, the *minifundistas*, lacking irrigation water and working at a low technological level. Coming to such holdings, as I often did, one was invited into the shade and offered refreshment, and one saw the many useful plants and the domestic animals underfoot. One noted the lack of creature comforts and the isolation too, of course, especially the difficult situation of the women, but the basics were there.

The ingenieros write of the economic decision-making in this "backward" rural world as though they were surprised, thus paying the incompletely commercialized and very incompletely mechanized or otherwise rationalized agriculture and ranching the strongest tribute. The producers, even the small producers, are shrewd enough not to cultivate what does not pay. If they need the money, they will sell, under protest, their hard-won sacks of grain or whatever for a pittance to anyone who will buy. They cultivate and nurture what they need to subsist, regardless of economics. They will be very careful of innovation, since their margin for error is very narrow. They resist what will affect net returns: the taxes and other charges, such as those that the ingenieros know will be needed to cover the contemplated improvements in the hydraulic works. Amazing.

Complementary Observations on One Ejido

Del Angel-Pérez studied the Vargas ejido on the eastern side of the Basin in 1989 and 1990 (1991, 1992). It is representative ecologically of the eastern side, since its lands cross the succession of microenvironments between the land subject to inundation and the dunes. She obtained several biographical sketches that go back to well before the SRH intervention and reviewed the historical literature; she found that not much had changed in ejidal production systems within the preceding several decades. Her main interest was the integration of the exploitation of plants and animals and the combination of that with outside wage labor. The results of the first two interests enlarge very well on the spare reports of the ingenieros; the third was not yet as important when the reports were written. By the late 1980s, about 45 percent of the Vargas ejidatarios would be employed in industry full- or part-time, mostly at a pipe mill just west of Veracruz city—TAMSA (Tubos y Acero de México, S.A.). Another 45 percent would be doing full- or part-time agricultural day labor on large holdings in the vicinity (Del Angel-Pérez 1991: 141–144).

The Production of Maize and Related Activities

Maize is grown for home consumption, for sale, and as feed for bovine cattle and household animals. Some maize is grown under irrigation, but mostly it is rainfed, or *temporal*. Ejidatarios prefer to cultivate it in riverine alluvium, the vega soils already discussed, but will grow it also on sandier soils upslope if that is the only option. Such production is precarious because it is not competitive with commercial maize production and because of the regional oversupply at the time when most ejidatarios are marketing their grain. At the same time it is very important; it is, after all, basic to subsistence in a direct sense. It is also very familiar, can be varied in location and timing, and requires little care and investment.

Maize is likely to be supplemented by beans for home consumption, and by some limited additional commercial crop production,

such as pineapple or *escoba* (*sorgo escobero*). Some ejidatarios on
the eastern side of the Basin are producing cane under contract
down on their best land, but this is more widespread on the northern and western sides. Some make charcoal for sale in neighboring
communities.

Del Angel-Pérez, with her background in an anthropology sensitive to ethnobotany, looks closely at what is around the houses—
the ingenieros' "disorder" (1991: 154–155, 205): she finds the full
array of fruit trees, spices and ornamentals, enclosed by live fences
of cacti or one or another of the trees out of the natural repertoire
of the surroundings. Underfoot are the fowl, the pigs—which serve
as a savings account, as sources of protein and means of exchange—
and, of course, the dogs. As already noted, there used to be a great
variety of fish as well as shrimp, turtles, and alligators in the waterways; there was game in the forest and ducks came down into the
aquatic vegetation. Most of these resources are gone, and useful
trees have been diminished by clearing. The ejidatarios, before they
were given land, themselves participated in this clearing. They
were allowed to cut wood on the private properties and make charcoal, which they sold in Veracruz, at the same time clearing the
owner's land for pasture. From what remains the ejidatarios bring
in wood for posts, for construction, and for fuel. She notes in passing that all this is managed on the basis of empirical, inherited
knowledge, not on institutional recommendations.

The Raising of Cattle

"Ranching" would give the wrong impression. The typical ejidal
"herd" is between ten and fifty animals. These are about two-thirds
criollo, or of the common mixed breed, and one-third zebu/suizo
crosses. Over one-half of the herd is made up of cows; the rest are
mostly calves, with a preponderance of females. The larger herds
are likely to include a bull, but service may also be obtained from
a neighbor.

The ejidatario is not likely to approach this activity with a capi-
talist mentality, we are told. Investment and returns are not as rationalized as one would expect in full-scale ranching on a substantial private property. Nevertheless, the milk, cheese, and meat that
may be realized can provide a tidy supplemental income. Since
there is usually very limited pasture available, animals must frequently be sold at the beginning of the dry season and, since many
people are in this situation, prices are low.

Also, Del Angel-Pérez detects a kind of empirical appreciation of the inefficiency of this transformation of energy. The animals are seen as being in competition with human beings, and hence it is not desirable to let the "herd" get too large (personal communication, 1991). The seasonality of the Basin and the environmental variation on its eastern side make transhumance desirable, but it must be managed with some ingenuity. The limitations of ejidal lands constrain the ambit, but movement can be facilitated through purchase or rental of additional parcels. Familial links may also play an important role. Relatives can take over from each other during outside employment or make room for each other's herds in times of scarcity. Thus *alto* and *bajo* lands are linked: hill pastures can be grazed during the wet season, and perhaps a bit of bottom-land in the dry season.

Risk may be diminished in one further way in the Vargas ejido, through the maintenance of a symbiosis between the ejidatarios and a private proprietor of substance (1991: 176–179, 209–210). This is a particular case and it is not clear to what extent such relationships prevail elsewhere, but it is remarkable on several counts.

The ejidatarios gain in various ways: the cattleman lends them money without interest and gives them advice. He buys their animals, mostly as calves, and fattens them up. He also buys their milk and transports it on to a pasteurization plant. In effect he is their *intermediario*. But he pays just prices and gives fair weight, they say. They prefer to do day labor for him rather than other neighboring proprietors if they can. He gains a stable labor supply and clientele; he also lowers the danger of the affectation of his property. They are happy to address him as *patrón* and speak of him to Del Angel-Pérez as a *cacique bueno* (personal communication, 1989).

Ranching on Medium- and Large-sized Private Properties

The ingenieros provide only a few basics regarding ranching beyond that practiced as an adjunct to cropping on smaller holdings, but many points of departure. What follows is an amalgam of their observations, our own, and those of others who have come this way in recent years.

A Day of Demonstration

On the 20th of October, 1961, there was a *demostración agrícola y ganadera* at the Central Veracruzan research station, Campo

Cotaxtla. The day's demonstrations were divided into two streams, one dealing with cattle raising and the other agriculture. I attended some of both and later obtained a set of statistics on the event (Table 7.5). It is interesting here for the way in which it set the cattlemen into relief.

The day of demonstration does not sustain the tardiness imputed to both agriculturalists and ranchers by the ingenieros. They are amazed to find in the 1970s that people living close to a major experimental station have not yet learned more about what has been found out there. Yet on this day, more than a decade before the reports are written, such learning is actually already underway. Both the organizers of the demonstration and the ingenieros have modernization in mind; this is the great necessity. But the latter are less sensitive to the economic context, to the needs and possibilities of their clientele, than the former; and the organizers have put persuasion to work, rather than just indictment (table 7.5).

Two worlds are represented here. In the one, ejidatarios and small holdings predominate; rainfed agriculture is the norm and the plow is drawn by oxen. But change is nascent; they are here at a day of demonstration, after all, and there are more than three times as many of them as there are those from the other world. In that other world private property predominates, holdings are far larger, and cattle are the main concern. New grasses have been widely introduced on these larger holdings. Irrigation, the ingenieros' main concern, is still definitely minimal in both worlds.

Appearance and behavior are as interesting as the statistics. The demonstrators, although they might have had the formal title of *agrónomo,* are kin to our ingenieros. They are in their element, giving instruction and advice. The sandaled ejidatarios, wearing the traditional four-dent jarocho straw hat, are quiet, awed, and observant. The cattlemen are booted, in pressed khakis and like as not wearing North American western-style hats, asking questions, offering comments, and driving off in pick-ups. It is not difficult to deduce what constitutes success and to sense old differences in well-being, in prestige and power.

The booklets available express something of the desiderata of the 1960s in rural Mexico. A young agrónomo/ingeniero is on the cover of one of them, "western" hat, pressed khakis, boots, pen poised over notebook. "How to harvest more beans in the tropics" is the title. Another has a grinning agriculturalist, in a jarocho hat but no longer wearing sandals, holding large cobs of hybrid corn. I can imagine him to be one of the minority of the "progressive" ejidatarios; I was introduced to many of them in the field. The title is

Table 7.5. *Excerpts from the Statistics of Participation in the Day of Demonstration (Mimeo).*

Percentages; NA = no answer, not applicable

Agricultural Stream
(296 registration questionnaires analyzed: 1/3 of total attendance in this stream)

Origin

Veracruz	87
Oaxaca	11
Other states	1
NA	1
	100

Occupation

Agriculturalists	85
Cattlemen	2
Agronomists	2
Teachers, students and others	7
NA	4
	100

Land tenure

Ejidatario	57
Ejidatario and private proprietor	9
Private proprietor	16
Renter, other	4
NA	14
	100

Total sizes of holdings

1.–4. hectares	27
4.–10. hectares	37
10.+ hectares	25
NA	11
	100

Irrigation

None	73
Some	9
NA	18
	100

Rainfed agriculture (*temporal*)
noted for virtually all holdings.

Traction

Oxen only	52
Tractors only	6
Combination of oxen, tractors and human	24
NA	18
	100

Cattle Raising Stream
(254 registration questionnaires: total of participants in this stream)

Origin

Veracruz	92
Other states	8
	100

Occupation

Cattlemen	89
Cattlemen and Agriculturalists	5
Agronomists	4
Others	2
	100

Land tenure

Proprietor	81
Ejidatario	8
Ejidatario and proprietor	2
Others	1
NA	8
	100

Total size of holding (hectares)

<10	7
11–25	8
26–50	11
51–100	23
101–200	20
201–500	18
501+	5
NA	8
	100

Irrigation

None	76
Some	11
NA	13
	100

Number of cattle

<10	6
11–25	15
26–50	22
51–100	24
101–200	11
201–500	9
501–1000	2
NA	11
	100

Additional information:

Cattlemen [see above] introducing new grasses	46
Participants in cattle raising stream with membership in some cattlemen's association	86

"How to harvest four tons of maize per hectare [a rather high yield]; seven sure steps to success." There are a number of booklets on the good grasses that can improve pastures immensely, and on ways to produce tropical silage that can overcome the scarcity of succulent feed in the dry season. Throughout there are repeated explanations of diseases, of pests, and of parasites—the scourges of the tropical lowlands, barely kept at bay.

The Cattlemen

Those whose properties and herds are in the middle range, are these still rancheros (Skerritt 1989a, 1989b, 1993)? Their independence, their involvement with cattle and horses, the predominance of private proprietorship among them, their assembling and disassembling of real estate, in these ways they do seem to fit that designation. They are certainly entrepreneurial. In other respects, such as the focus on the family for labor, the relationship is not so clear from our evidence, and even less apparent are the strict control over family and the religiosity that have been attributed to the ranchero. The figure has had its own dynamics, as Skerritt points out (1993: 179).

Those at the upper end of the scale, the cattlemen of substance, are reminiscent as much of *hacendado* as ranchero. Their hospitality is unstinting: cool drinks on verandahs and open invitations to perforate their wetlands in the pursuit of prehistory, old world graciousness with a cellular phone. These men are current and show style; they go to the World Series baseball games, keep race horses, or raise fighting cocks. They are like Sartorius's working landowner friends who take newspapers, have contacts in the cities, and are men of influence. Authoritative? No doubt. Also beneficent? Perhaps. Del Angel-Pérez certainly found it so in the community she studied (1991: 128).

An important issue in Southern Veracruz in the 1960s was the invasion of private lands by those with no land or not enough. Several times we came into localities just before or just after shootouts between parties in such disputes, and I must say that the naive Canadian graduate student rather regretted having missed the excitement. My guide during much of this fieldwork was an agronomist with good contacts to ejidatarios and the people in the new colonies, more associated with the have-nots than the haves and thus vulnerable to the landowners' hired gunmen. In any case, he tended to wear an automatic under his shirt. Occasionally we would see army contingents on the move to dislodge invaders. Cattlemen took out full-page ads in regional and capital newspapers protesting the insecurity. How could they produce what the country needed under these circumstances? In their reports the ingenieros just indicate that there is little of this in the Basin in the 1970s, an important positive point for the decision-makers, whether fully justified or not.

On different occasions, in different parts of lowland Central Veracruz, I elicited, without trying, muted regret from small holders, especially ejidatarios, that they do not have access to the lower humid land in the Basin, the land that once sustained intensive wetland agriculture. If they did, they would be able to grow dry-season crops and develop larger herds.

The Animals

The ingenieros tell us that in 1973 there were 163,325 bovine cattle in the six *municipios* from which the two irrigation districts are carved (SRH 1973: 113). That is not far from the figure of 150,000 given by the médico in 1580 for cattle in a district of the Basin with a radius of seven leagues, which in turn is not too different from the area enclosed by the six *municipios* (Figure 7.3). There were 37,802 bovine cattle in 1973 in the two districts, Actópan and La Antigua, together (table 7.6). We know that most of those were

Table 7.6. *Animals in 1973 (Source: SRH 1973: 114).*

	Total	Pure	Crossed	Criollo
Total Nos. for Two Irrigation Districts				
Bovine				
Milk	409	5 (1.2%)	375 (91.7%)	29 (7.1%)
Beef	37,393	507 (1.4%)	33,092 (88.5%)	3,794 (10.1%)
Total	37,802			
Porcine	2,555	78 (3%)	783 (30.7%)	1,694 (66.3%)
Equine	3,543	0	834 (23.5%)	2,709 (76.5%)
Ovine & Caprine	988	203 (20.5%)	0	785 (79.5%)
Avian	15,274	2,998 (19.6%)	2,626 (17.2%)	9,650 (63.2%)
No. of Animals on Private Properties in the Two Districts				
Bovine				
Milk	308	0	279 (90.6%)	29 (9.4%)
Beef	31,265	250 (0.8%)	28,947 (92.6%)	2,068 (6.6%)
Total	31,573			
Porcine	1,083	20 (1.8%)	681 (62.8%)	382 (35.2%)
Equine	1,926	0	533 (27.7%)	1,393 (72.3%)
Ovine & Caprine	587	170 (29%)	0	417 (71%)
Avian	6,533	644 (9.9%)	56 (0.8%)	5,833 (89.3%)
No. of Animals on *Ejidos* in the Two Districts				
Bovine				
Milk	101	5 (5.5%)	96 (95%)	0
Beef	6,128	257 (4.2%)	4,145 (67.8%)	1,726 (28.2%)
Total	6,229			
Porcine	1,471	58 (4%)	101 (6.9%)	1,312 (89.1%)
Equine	1,617	0	301 (18.6%)	1,316 (81.4%)
Ovine & Caprine	402	34 (8%)	0	368 (92%)
Avian	8,742	2,354 (26.9%)	2,571 (29.4%)	3,817 (43.7%)

in the latter, and that is about as precise as one can get for the Basin. Of the total in the two districts, 31,573 were on private properties and 6,229 on ejidos.

This intriguing parade, articulate to the last digit and complete with decimal points, invites scrutiny. The dichotomy of private property and ejido is not as sharp in this case as it may seem. The first category includes a range of sizes; the private properties on the small end are not too different from ejidal holdings in nature and purpose; indeed, some rural people have both ejidal plots and small private properties. The importance of beef and the crossed breed are indicated, whereas that of the double-purpose herd is hidden. Many beef animals are milked, many old cows sold for meat, we know that. The pig is a savings account, here as elsewhere in the developing world; it is more important on ejidos than on private properties. The small herds on ejidos provide some additional income and represent savings, too. "Equine" means horses, donkeys, and mules, but here mainly the first two; not many mules are needed now, except perhaps on the larger ranches for the transport of cans of milk. On the ejidos, the figure must represent mostly the donkey, the old burden-bearer for the humble household. There are few sheep and goats in the lowlands, as always. For "avian" one must read chickens, ducks, turkeys, and geese—there are quite a few more on the ejidos than on private properties. One can hear the noises around the houses, and the chop in the backyard that means a meal is being prepared for the visitor. The animal missing from the list, but always there, is the dog.

The most desirable animal for ranch stock is the cross between Indian and European strains. It is the main means of herd improvement and expansion in the tropical lowlands; it makes the development of herds for both meat and milk possible. Any rancher asked about breeds and yields in the 1960s, 1970s, or 1980s said roughly the same thing (Rios 1985).

A representative of a prominent Veracruzan ranching family was interviewed in a Veracruzan newspaper in June of 1985; they were credited with having introduced cattle of Indian origin, commonly known as zebu, into the state in 1923 (Rios 1985). There is independent confirmation of the date (Cisneros et al. 1993: 223). The hump, the long rippling dewlap, the bulging forehead, the sloping ears, and the light gray color grading into browns add up to an easily recognizable phenotype. It is subdividable, of course; there are several common varieties, including Gyr, Nelore, Guzerat, and Indobrazil. The last is interesting because it indicates something of the route by which the cattle reached Veracruz. The zebu is noted for its resistance to the rigors of the tropical environment, especially the high temperatures. While other breeds pant in the heat, this animal breathes normally and does not become anxious; it obviously has a very effective way of regulating its temperature.

Suizo is the Mexican term for the European breed or breeds noted for milk production and most often crossed with zebu. The combination of hardiness and such productivity has turned out to be ideal. The proportions of these two strains, the Indian and the European, often with a residue still of the common mixed breed of the land, the *criollo,* will vary of course between and within herds, reflecting the degree to which the particular owner has been able to upgrade his herd.

Cattle dominate rural change in Central Veracruz in the 1960s; they are the prime means of economic ascendance. Cities are growing and people are living better and better; the country needs meat, say the politicians, who like as not are cattlemen themselves. A similar spirit of promotion is shown to pervade the ranching of Tabasco during the early 1970s (Barbosa-Ramírez 1993); it was, in fact, quite general throughout the country.

Our own informants were always proud to show us the bulls with which they were upgrading their herds (figure 7.6). These might still be multicolored, -sized and -shaped but would soon become more uniform, less *criollo.* The cattlemen on the make were often still spraying the beasts against the tick, but they planned to have a proper dip chute soon. We could see the forest retreating. In many pastures there were still its remnants—live trees left for shade and those already dead but not yet down.

The expansion of cattle ranching in Veracruz traces a straight, steep line from 1940 through to 1980. During that same time, state production is a greater and greater percentage of national production until 1960, when the percentage drops off; other parts of the country are gaining on it (Cisneros et al. 1993: 213–214.)

The Grasses

The extent of the grazing land in the Basin is unclear because of the necessary overlap in the land-use categories that the ingenieros try to use. One of them observes cogently that the area that can be considered to sustain cattle is not only natural grassland and planted or seeded pastures but also, since cattle graze and browse, patches of low forest, or *monte,* coming in during a fallow period,

Fig. 7.6. (*a*) Roundup of cattle on a medium-sized lowland Veracruzan ranch in 1961. The animals are largely *criollo,* with admixtures of European and Indian breeds. An animal is fastened to the central post of the corral and is being sprayed for ticks. The pasture in the background is African grasses planted on recently cleared land. (*b*) The owner of the ranch posing with his recently acquired zebu bull.

as well as croplands that are grazed for their detritus (*rastrojo*) after a harvest, and the borders of canals and fields (SRH 1973: 126). One could add the feed provided around the household and what is cut elsewhere and brought in to feed confined animals. The determination of the sustaining area for cattle is further complicated by seasonality. Some prime grazing land, for example, is annually inundated, but the extent of the inundation varies from year to year. Under these circumstances, any calculation of carrying capacity is illusory.

African grasses are the most important sustenance of Central Veracruzan cattle. They were introduced into the Americas in the second half of the nineteenth century and on into the twentieth century, as James Parsons showed us (Parsons 1989: 299). They are used on "better" drained soils and do well under irrigation. These are lush, high grasses mostly, within which cattle disappear, to be called out again by the vaqueros. Such pastures are achieved by seeding or the planting of cuttings, and here Del Angel-Pérez has an interesting footnote. To do this work around Vargas, children are hired: they are closer to the ground, do not complain of back pains, and just keep going (1992: 21–22).

Native grasses on the other hand, although less palatable and productive than introduced grasses, are used in areas subject to "poor" drainage or inundation. Fairly good pastures of native grasses can be "induced" by removing woody growth. Such pastures on the more humid soils, particularly in the bottomlands proper, do not usually dry out seasonally and hence do not need yearly burning to remove old growth in the same way as the pastures, or "savannas" on the hill land to the west.

Ranching Practice

The material culture of ranching, which the ingenieros regard as antiquated, is morphologically interesting. Many of the artifacts and related terminology date back into the colonial period. Some are recent innovations or adaptations to this environment. Fencing is done mostly with barbed wire; where the posts are tree trunks out of nearby forest thickets, they are likely to sprout and form live fences, one of the striking features of this landscape. Where it can be afforded, cement posts are used instead. Watering facilities may be improvised from old bath tubs and petroleum barrels or they may be well designed, covered troughs, but they are usually carefully placed between pastures. There are also watering systems on undulating terrain that feed water by gravity from a central tank to float-activated troughs in the various pastures. Pens, gates, and walkways facilitate grouping and isolation. There is usually a brutal contraption in a narrowing of one of the walkways where an animal can be stopped and immobilized with a hinged grill for whatever needs to be done to it. In the assemblage, one is likely, now, to see a long, narrow dipping tank and a loading chute. In the center of the corral there is usually a forked post, perhaps a topped tree left in place, to which lassoed animals can be tied. The equipment for milking, for cooling and transporting the milk, is mostly rough and made by hand of wood, metal, and rope. It does not encourage one to drink this milk unboiled, but it is all very serviceable under the circumstances. The covered milking place, the *ordeña*, is usually on some rise; underfoot, as we have said, there are often sherds.

On many ranches the main thing is the fattening of young stock. These animals are probably obtained as calves, perhaps from the ejidatarios with limited pastures, and then brought into the bottomlands for fattening. On the most advanced ranches, the emphasis is likely to be on raising breeding animals from registered stock; this is often combined with the fattening of young stock for beef, and with milking, as well. These are the ranches that enter animals in fairs. Their brands are seen as logos on pickups; ribbons and photographs of prize animals hang in their offices. The more modest operation is likely to be simply a dual-purpose beef and milk operation.

A very prominent Central Veracruzan cattleman remarked to me one day several years ago, apologetically, that the ranching of the region was still quite rudimentary. Perhaps it was when measured against the large dairies and beef feeder lots of North America or, indeed, northern Mexico. The same informant explained to Del Angel-Pérez, however, that full-scale stabling and controlled feeding, such as the ingenieros consider modern, would be uneconomical in the Central Veracruzan context (1991: 105). She takes a long view and maintains that technologically the large commercial ranch is quite modern in some respects, especially in insemination and health care, but traditional in others, in fact much like ranching was a century ago, and for good reasons (1991: 124, 132).

The ingenieros are very critical of animal health care on the ranches they have seen (SRH 1973: 141). The decade or two between them and Del Angel-Pérez seem to have made a difference. My own inquiries in the 1980s indicated that ranch foremen and the vaqueros, too, carry a good fund of bovine folk medicine, as well as a rudimentary knowledge of how to apply modern remedies.

Transhumance

The ingenieros mention this movement only briefly (SRH 1976a: 261), but for a moment one senses again the seasonal beat of the landscape and is reminded of the ecology of ranching. Transhumance has a very restricted ambit in ejidal ranching and on small private properties, as has been noted, but on larger private properties it involves major shifts. The animals are moved seasonally between environments; they are taken out of the wetlands to pastures in neighboring hill land at the onset of the rains and the beginning of flooding, then brought down again with the onset of the dry season when water levels in the wetlands drop, when the vegetation on surrounding hill land is graying but that in the bottomlands remains lush.

On ranches with contiguous holdings in various microenvironments, upslope and downslope movement is simple; it is facilitated by fenced walkways. The placement of the drinking troughs and the related wells and fences on long gentle slopes is often diagnostic of these seasonal shifts within the landscape (figure 1.11). We found repeatedly that wells were located near the maximum flood level and could be reached from below during the months of grazing in the bottomlands, as well as from above during the time of the rains. In other words, they were placed on the boundary between zone *a* and zone *b* of our heuristic model (figure 1.9).

Transhumance happens between many of the larger properties. Del Angel-Pérez observes that the same family names come up frequently among owners of ranches in the various environments of Central Veracruz. A given owner may thus have properties at various altitudinal levels or be able to make arrangements with relatives. Strategic marriages have allowed the aggregation of some lands and thereby facilitated transhumance (1991: 104, 119, 133–134).

Wet-season pasturing upslope plus the transport back and forth is the stressful time for the cattle. The months in the bottomlands, usually the greater part of the calendar year, are the time to fatten.

I had a most agreeable discussion one late afternoon with a ranch foreman surrounded by vaqueros. He was very much the authority amongst his men, and richly knowledgeable, quite like the foreman Sartorius talks about. I asked, among many other things, how they got the cattle out of the bottomlands after the onset of rains without any of them being marooned by floodwaters, which is a common problem on ranches in the lowlands. Floodwaters can rise a metre or more within a matter of hours. Well, they had experience, he said. I could imagine the detailed knowledge of the topography, and other lore as well, that would be necessary to move animals quickly under these circumstances.

Cattlemen have altered the wetland environment. Some have cut herringboned networks of drains across their wetlands and over the remains of Prehispanic platforms and canals, in order to hasten the descent of the floodwaters toward the end of the rainy season and, with that, the entry into renewed pastures. Some have drilled wells or cut intake trenches into the lowest parts of the wetland and installed spray irrigation systems to serve the upper wetland margins and adjacent terra firma during the dry season. In various places within the Central Veracruzan bottomlands, we have noted recent construction of stone-lined causeways over particularly low areas in order to facilitate the movement of animals, as well as pickups and machinery. They certainly made our own access to test excavation sites easier.

Recapitulation

The efforts of technically trained people to take stock of conditions in the Basin, to gauge the impact of the hydraulic intervention of the 1940s and 1950s, to prescribe for a revitalization that did not happen, proved very helpful to our project. We gained excellent air photos, large-scale maps, environmental measurements, and numerous other observations that have been most useful for the investigation of Prehispanic wetland agriculture around A.D. 500 and for the treatment of subsequent epochs. Taken with other sources and our own field observations, the SRH reports made it possible to make a final cut in the analysis of the history of land use in this one diagnostic region, to pursue into one further epoch the theme of the persistence of certain tenure patterns and basal agriculture. The bureaucratic materials, though, are unamenable in method and perspective.

The ingenieros present us with their own geography of favor. They see the Basin and its immediate surroundings as a poverty-stricken area, not too different from many others in Mexico, but with the potential for an expansion of irrigation and thus intensification. The most hopeful, or the least impeded, lands for them are toward the lower end of our zone *a* (Figure 1.9), particularly the gentle alluvial slopes above usual flood levels. Here is where commercial agriculture has long been carried out and where it can be enhanced. The lesser agricultural lands extend upslope from that, impeded substantially in various respects; with sufficient investment their use can be intensified. Much leveling and clearing of stones would be necessary, as well as careful cultivation practices to guard against erosion, all of which is expensive.

To the extent that the Basin can eventually be further drained and then served by irrigation canals, the favor can be brought downslope through zone *b* and even into zone *c*. Mostly, however, the swamps and the lagoon of zone *c* are negative entities in the ingenieros' evaluations. The cattlemen's geography of favor is thus obscured. For the proprietors of substantial holdings, as well as the ejidata- rios and other small holders struggling to establish small herds, the lands of zone *c*, especially its upper margins, are of great importance already; in fact, they are key to their operations. This is different from the assessment of the modernizing technocrats. For them, these are swamps that must be drained. The ranchers' views, the historic evaluation in these parts, is akin to the ancient agricultur- alists' appreciation of the land subject to inundation.

This last rendition, too, can end in the pastures. There are no more *cimarrones*; there are no more places for them to hide. The hump-backed and many-hued beasts of a modern herd are still sometimes driven up- or down-slope, but more likely are trucked. The detail of the ecology of ranching engages the energy and inge- nuity of the cattlemen and their people as before, but the process and the metaphor of transhumance has faded with the efforts of the ganaderos and the ingenieros to mitigate the need for movement with the seasons.

Postscript

The modernization envisaged and promoted by the ingenieros and agrónomos has graded into what may be called globalization or transnationalization. The ambit of Mexican trade has widened; its constraints are being progressively dismantled, or "liberalized." This process now overshadows the Basin and the whole of rural Mexico. The legal terms of reference of tenure and the direction of governmental involvement in the economy have changed. The op- tions in both modernized commercial and traditional production have narrowed as comparative advantage becomes the greater good. It is highly likely that ranching on the larger properties will adapt and persist. Traditional production is faced with the challenge of adaptation, and many other consequences raise concern as well.

The examination of the Prehispanic agricultural system in the wetlands and what has succeeded it, especially the basal agricul- ture practiced alongside other pursuits from early colonial times to the recent past, all this is highly relevant to the pathology of the agrarian present. It can be taken as a basis for the progression of conceptualization into the 1990s. The consideration of the new winds of change, the masses of absorbing new literature on rural Mexico and the actual changes in rural communities, including those around the Basin itself, is best undertaken in another framework (Siemens and Heimo 1996).

8. Summing Up the Yields

"No description without performance."

—J. Brian Harley 1989: 2

Wherever one looks in Central Veracruz there are remains of ancient occupance; scan a horizon and the natural roll of the topography is likely to be interrupted by mounds. Still, the paleoecology and cultural development of the region are relatively sparsely known and its chronology has been considered "soft." It was there at an important colonial beginning, and yet many aspects of its history have only lately come under intensive examination. This analysis has lifted out one of its lowland basins, finding it coherent, instructive, and, indeed, representative in both physical fabric and human imprint.

Substantive Additions

We, the collaborators, have sensed the Basin's seasonal pulse, responded to the signals of its vegetation, pitted and cored its stratigraphy with a will. We have appropriated what we could from many sources and our own observation, and then made the gentle indentation visible by a deliberate choice of perspective and graphic exaggeration. We came to appreciate one particular organizing principle.

Verticality

Humboldt outlined the general altitudinal scheme that is now commonly taken for granted in the explanation of climate and vegetation in tropical regions with high relief; its main variable is temperature. Sartorius elaborated on it for Central Mexico. Subsequently Troll (1959), Murra (1978), and many others in recent decades have worked out its physical detail and cultural implications in the Andes, with frequent comparative observations regarding other tropical mountain regions. We have tried to bring some of that to bear on Central Veracruz. The remains of the Prehispanic planting platforms and canals occur on the lower extremity of the scheme. Here the scale must be enlarged to accommodate very limited relief, and the emphasis must be on the relationship of micro-topography to water levels.

Whatever the scale, vertical variations are best shown in schematic profiles. West and his colleagues (1969), as well as various botanists, including Gómez-Pompa (1980), have generated helpful symbolism in this regard. The device actually demands a good deal of creativity and is not easy to master. It must simplify but remain believable. Around wetlands a great deal can be arrayed along one gentle slope.

Any particular arrangement of natural features and productive adaptations is likely to prove dynamic, and precarious. Altitudinally interdigitated or juxtaposed microenvironments are often exploited in a complementary fashion. Orchestration, the cueing in of activities in these various settings, sometimes emphasizing this and sometimes that, is the key to effective production. It combines a changing complement of activities into at least a potentially balanced, ecologically sensible, and resilient yearly round. Diversity and complementarity is the advice that such a system holds up to the juggernaut of globalization.

Natural hazards, of course, may affect the combinations in any given year. Hurricanes menace the Gulf coast from time to time; every decade or so there is the possibility of a much higher than normal flood that will destroy what is usually safe. *Nortes* come every year, sometimes damaging dry-season crops on the precious moister low-lying lands, thus vitiating the wetland advantage. Then there are the uncertainties born of annual variation in timing and totals of precipitation, to say nothing of secular changes, which we are only beginning to understand.

The altitudinal scheme, especially when elaborated at a large scale around its lower end, is a fine heuristic device. It helps to organize contemporary observation, to reformat heterogeneous compendia in the documentary sources, and to articulate the Prehispanic wetland agricultural system.

Raised Fields

Wetland agriculture has now been of continuing academic interest for some decades. Charles Stanish has restated confidently what one or another of us might have ventured in unguarded moments some time ago: "It is not simple hyperbole to state that raised-field agriculture is among the most important intensification strategies utilized by Prehispanic farmers in the Americas" (1994: 312). His data and argument substantiate and elaborate on the contention rather well. Andrew Sluyter has done all of us a service with a comprehensive review and an upbeat assessment of Mesoamerican research on the subject, as well as a rich original contribution concerning Central Veracruz (1994, 1995). B. L. Turner II has repeatedly and incisively reviewed the theoretical issues surrounding the investigation of wetland agriculture in Mesoamerica (1985, 1993). It is very much an ongoing concern.

Those involved are likely to agree that improvement of the data base has been an urgent necessity from the beginning of the investigation of patterning in Mesoamerican wetlands, and it remains so. This study, the sequels in process, and those planned for the future are intended to enhance the base, to contribute to the refinement of the understanding of the function, chronology, and significance of the system. It has also seemed important to push the discussion progressively further out of the Maya realm, where the first Mesoamerican remains of the system were found but with which it is unduly associated.

The San Juan Basin investigation has provided a clear example of Prehispanic wetland maize cropping. A relative chronology has been proposed: the Early Classic, or ca. A.D. 500. (Siemens et al. 1988: 107). This has been firmed up by independent direct evidence, particularly Sluyter's core out of Laguna Catarina (1995: 202, 236). We have modeled agricultural practice and change along wetland margins and reconstructed activity in the Basin as one might have seen it in the time of use.

The scene reflects several basic hypotheses regarding function: the patterning in the wetlands of the Mesoamerican lowlands mostly represents *proto-chinampas*, not *chinampas*; calling the system simply *planting platforms and canals* is a good way to avoid tenuous associations. Water levels, it seems fairly clear from a close study of the morphology and sedimentary context of the remains, were not controlled to the extent that year-round cropping could be carried out. Instead, use was probably seasonal, interphased with, and benefiting from, fairly regular flooding. Agriculture on the platforms was probably preceded, or followed, or even accompanied on neighboring terrain, by fugitive, flood-recessional agriculture.

Wetland agriculture on the planting platforms is often termed intensive (e.g., Sluyter 1994), and that serves in a general way, but such agriculture was itself subject to intensification and disintensification. Initially flood-recessional agriculture on unmodified floodplain surfaces will have accommodated some of the prosaic yearly variations in requirements and absorbed some of the shock of misfortune or special need.

A wetland near El Palmar provided us with repeated object lessons in this regard (Siemens 1989a). In some years during our prolonged period of observation, the use of land subject to inundation was considered worthwhile, in others years not. One major variable in this case was the increase and decrease of the availability of nearby industrial employment. Wage labor often provided what was needed beyond what the annual wet-season cropping of hill land could provide. One can imagine a whole series of historic and prehistoric factors that would have had similar effects.

In the early centuries A.D. it is likely that larger imperatives will have overtaken the well-meshed yearly round from time to time; tyranny and conflict will have overshadowed it again and again. It clearly became necessary sometime in the Classic period to intensify flood-recessional agriculture or, one might say, to increase the investment of means and effort through canalization and the building up of planting platforms, probably beginning at the upper margins of the zone of annual inundation and then pushing successively farther downslope. For this we do not have contemporary or historical agricultural analogues in the lowlands. Eventually factors about which we can still only speculate will have intervened to make it unnecessary or impossible: social disintegration, strategic relocation of settlements, or increases in sedimentation due to deforestation upslope and resulting changes in soil conditions and in average maximum and minimum water levels. Long-term environmental change may well have been involved; this is being given increased attention (e.g., Gunn et al. 1995; Hodell et al. 1995).

In any case, it seems useful always to think of wetland agriculture in the context of other productive activities. It is likely to have taken its importance less from its function *per se,* than from its role as a flexible element in an orchestrated exploitation of adjacent microenvironments by a variety of techniques.

The history of ranching in the lowlands echoes some of these agricultural dynamics along the hypothetical gentle slope. Feral cattle sought out the greenery of the wetland margins on their own during the dry seasons—roughly paralleling ancient incursions for hunting, for fishing, or, indeed, for fugitive seasonal agriculture. Ranching in the wetlands and their surroundings has for some time been one of the main means of achieving prosperity. Its intensification in the late-nineteenth and twentieth centuries has largely tended downslope; by the cutting of drainage canals, water levels can be brought down more rapidly after the rains and to lower levels, thus lengthening the annual access to areas subject to inundation—in effect, bringing down the boundary between zone *b* and *c* in our model. Bridges and even causeways are built within pastures to facilitate access. Pump irrigation systems are installed to irrigate pastures on the upper portions of the zone subject to inundation, once the waters have receded and the pastures have become dry, or to irrigate and enhance pastures on firm ground, in zone a. The ranchers have thus altered the wetlands quite aggressively. Their methods are still commonly labeled "extensive" by analysts and by the ranchers themselves, but the designation is relative, involving a comparison with beef-feeder operations.

There is little evidence from colonial times, and certainly not recently, of the disintensification of ranching. However, there is that interesting observation by Corral about the ephemeral nature of ranching in any given place, implying that it may have waxed and waned considerably in historic times—a proposition that needs more study. It is difficult for us now to imagine a major lapse of ranching in the lowlands.

Our interpretation of patterning in the San Juan Basin pushed against the envelope in regard to the morphology of wetland agriculture. We did not attempt to plot the platforms and canals in the Basin by circumscribing eroded remnants, but by using canal midlines. We mapped to a detail and an extent not matched by the investigators of other locations. We detected and offered a rationale for a frequent variation in form from the margins of the wetlands with terra firma down to the swamp proper or the lake—elongated and parallel at the top and quadrate at the bottom. This was taken to indicate not only a sequence in the initial construction of the system, as outlined, but also variations in land and water management through a normal dry season: efficient drainage to allow rapid entry at the outset and baffles to retain water and thus facilitate both movement in the canals and subsurface and scoop irrigation during the critical final weeks of the dry season. The distinction has gained some acceptance. Less plausible, evidently, are the discontinuities in patterning that hint at variation in tenure or phasing in construction. Least acceptable of all, although it is one of the most striking and consistent characteristics of wetland field complexes throughout Central Veracruz, is the evidence for an orientational gist, which fairly invites a cosmographic inference.

Sluyter has assiduously assembled the basic locational and chronological information available on Prehispanic settlement phases in Central Veracruz (1995: 143–153, 457–475). Systematic field surveys and a great deal more would be necessary, however, for satisfying theorization regarding cultural ascent and descent on a small scale. Nevertheless, we have seen and deduced enough in the Mesoamerican lowlands to question the grand scheme, the "meta-narrative" once considered relevant. Stanish puts it well: "The simplistic explanations of the hydraulic hypothesis remain untenable. . . . more subtle causal relationships exist between political centralization and agricultural intensification" (1994: 312).

Our detailed examination of a prehistoric wetland agricultural system in one place, complete with semantic quibbles and qualifications, has been geared precisely to suggest the kinds of functional subtleties that are needed for an elaboration of the larger theoretical context.

Historical Overburden

The Prehispanic patterning has its overlays. Rather than just peeling them back, we chose to examine them in turn, probing for continuities as well as discontinuities, noting how the Basin was used under one set of terms of reference and then others. This bore down on such successive concentrations of coherent materials as could be found for Central Veracruz. The discussion often arced to affect what came before and after.

Contact brought the Indian population in the lowlands a decrease of some 90 percent. This had its ecological impact, of course, in Central Veracruz (Sluyter 1995), as elsewhere. In the discussion of this impact it is important to qualify the sweeping, quotable small-scale observations with distinctions between microenvironments (Siemens 1995a: 207–209).

The historic materials are in various ways more difficult to deal with than the prehistoric, but reconstruction of the early colonial lowland Central Veracruz was attempted nevertheless—a synthesis rather than just an extraction of the diagnostic, as has been usual. This world is presented to us from Old Veracruz. The documents' many, many "bytes" are reformatted to achieve a modicum of ecology. The Prehispanic elements of that colonial landscape are faint, the Indians themselves are few and distant, agriculture is hardly apparent and ranching is obtrusive. In the foreground is the town; it is given the preeminence one would expect in Spanish descriptions. Freight is being transhipped and carted, muletrains are leaving and returning. Across the river are the lush, difficult wetlands and, some leagues away, the *estancias* of the Ruízes.

Corral informs us about the Veracruzan lowlands in the late eighteenth century. His material on Southern Veracruz is focused, a kind of geography of wood and transportation in aid of a proposed shipyard. The portion on Central Veracruz is an afterthought, but we do get an update on ranching—and an intriguing map.

For the middle of the following century we have a description of Central Veracruz that throws long shadows. Much of the world that Sartorius presents is still colonial, especially with respect to ranching. The *estancias* have merged into cattle haciendas, but ranchos are apparent too. Sugar processing is being modernized in places, colonization and new types of commercial cropping are envisaged, but most actual agriculture is shown as still very rudimentary.

For the first time we have environmental descriptions for our region that can be taken fairly straightforwardly, and they remain explicitly instructive in some aspects to this day. We are taken across the northern part of the San Juan Basin, which is critical for this analysis. The savannas and the *barrancas* are particularly impressive, and the whole is vertically articulated.

Then we come on the SRH's magnificent, large-scale, late-twentieth-century air photographs and topographic maps with contour intervals of one metre, expressly for the Basin! These and the massive reports provide very useful environmental information for the contextualizing of the prehistoric remains and all that came after. We are given details on the massive latter-day engineering incursion into the Basin, in aid of the intensification of agriculture. Inevitably one compares it with the much earlier incursion.

In all of this one can find a great deal about recent land use, especially when supplemented with the findings of Del Angel-Pérez (1991, 1992) and our own fieldwork. The ingenieros' own materials are not terribly coherent—or congenial. Their world is recent but not quite contemporary. The agency's intervention in the Basin lays an engineered hydrography over a network of arroyos and meandering streams, a "text" overwritten.

A Tenacious Succession

From this diachronic analysis, two strong tendencies have become clear. Each is a regional variant on larger phenomena, and each touches on some contemporary nerves. The first becomes apparent when Prehispanic patterning in the wetlands of the San Juan Basin is superimposed on modern land tenure (figure 8.1). Lively commerce in real estate, bureaucratic inertia, and who can say what other factors have made a fiction of many of even the latest and most official Central Veracruzan cadastral maps. However, our own fieldwork shows that the maps can be credited generally with

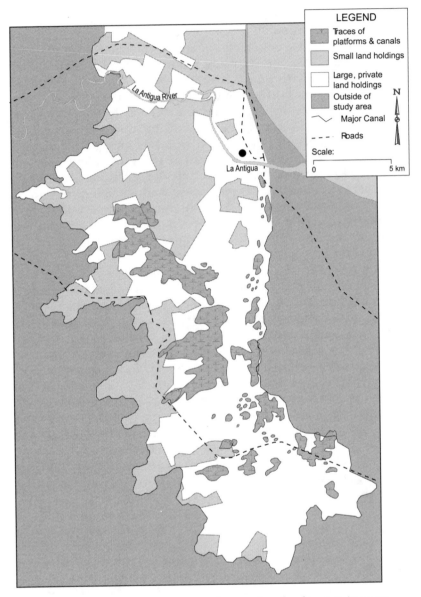

LEGEND

Traces of platforms & canals

Small land holdings

Large, private land holdings

Outside of study area

Major Canal

Roads

Scale:
0   5 km

N

La Antigua River

La Antigua

Fig. 8.1. Prehispanic patterning superimposed on land tenure in 1970.

respect to property sizes even though any particular boundary or assignation of ownership may be questionable. The striking basic point is that ancient remains are located predominantly on the larger, contemporary private properties, which in this area are mostly ranches.

The correlation brings to mind that the heritage of transhumant ranching in the riverine lowlands of southwestern Spain was initially applied within expansive holdings in the Mexican Gulf Lowlands that included wetlands, and carried to extremes by feral animals. The nineteenth-century parcelization of haciendas dispersed but did not vitiate the basic advantage of access to wetlands in the dry season. There were changes in the nature of ranching. Haciendas became ranchos—different in size and manner of production, modernizing as the twentieth century wore on (Skerritt Gardner 1989a). And yet, as Del Angel-Pérez found on the eastern side of the Basin, some larger properties still have the echoes of earlier times; she sees *neo-latifundismo* (1994: 218). Agrarian reform was not able to dislodge the people of substance, or in any case did not do so in this region; ejidos surround the wetlands and ejidatarios speak now of wanting them and what they could do if they had them, but mostly these lands remain in private hands. Ariel de Vidas recognizes something similar in Northern Veracruz; it is a "sleeping beauty": agrarian movements, the petroleum boom, and much else has passed over the landscape, but its basic colonial fabric, the *feudo ganadero*, remains immutable (1994: 39, 41, 68).

Thus, land that was once used for agriculture that we regard as relatively intensive is now used relatively extensively. The vestiges of ancient agriculture have been made much more visible by forest clearance, but the favored terrain they occupy remains inaccessible to those who would like to use it for agriculture. Propositions for the reactivation of the ancient system in aid of improved subsistence for presently marginalized rural people come squarely up against this basic fact.

A Continuous Undertone

The main economic pursuits of Central Veracruz—through-trade, ranching, and commercial agriculture, especially the cultivation of sugarcane—appear repeatedly in our historical materials. There is another sort of productive activity as well, involving a multiplicity of plants, animals, and techniques and the integrated use of neighboring microenvironments. It is directed toward subsistence and the gaining of a limited surplus and is detectable underneath or alongside the various main pursuits. We may deduce it for prehistoric times in Central Veracruz and we have seen it in the recent past.

Such contextual activities are difficult to name and define. One of the best characterizations runs as follows:

> [They have] remote origins, are ecologically adapted, sustainable over long periods of time, and use large amounts of human labor rather than agricultural machinery. [They are] apparently technologically very simple but are, in fact, extremely complex in holistic aspects. They take into account the different elements of the local ecosystems in which they are inserted. Agricultural knowledge is transmitted informally from generation to generation. These systems are also very flexible in their adaptation to market conditions. (González Jácome 1993: 142)

Such systems are often referred to uneasily as "traditional" agriculture. Some useful alternative terms are available, such as *agricultura básica* or *autoabasto* (Skerritt Gardner 1989a: 8). Such systems also have been called a "Mexican model," following Angel Palerm, the much respected scholar of the rural Mexico of a few decades ago. The model is counterpoised to practices adapted to extraneous conditions and organized according to economic, not ecological, principles (González Jácome 1993: 142). Modern, mechanized, and globalizing agriculture is opposite to "Mexican," "basic," or "traditional" agriculture in virtually every respect.

Unease over the word *traditional* has various theoretical and practical roots. Max Weber, in his outline of types of social action, distinguished traditional from rational and indicated the first as being immune to the second (Parsons 1937 [1949]: 645–647). One might not want to look closer at a phenomenon thus defined or at least evade the use of a term that stigmatized it. *Traditional* has often been given other negative or ethnocentric connotations. Agriculture so qualified may well be considered resistant to change, obsolete, undeveloped, and certainly not modern. This judgment has often been made in Western society with respect to various aspects of non-Western societies (González Jácome 1993: 141). Nevertheless, the term is still the one that first comes to mind; it is almost unavoidable, whether flagged as doubtful by quotation marks or not.

Virtually every one of the characteristics of the "traditional" noted by González Jácome can easily be construed as highly advantageous. And, it is interesting to note as well that Weber, when one looks a little further, not only considers the "traditional" as immune to the rational, but as action that is sanctified and legitimized (Parsons 1937 [1949]: 659–660). The word is thus a good deal more than a designation by default.

"Traditional" is, on the face of it, an historical concept, but it can be projected into Mesoamerican prehistory. It is difficult to use an antonym like "commercial" in that context, but it is quite likely that something similar to the dichotomy between production of a surplus and the necessities for subsistence also obtained. In any case, the agriculture that can be imputed to the planting platforms, and indeed to the antecedent flood-recessional agriculture too, together with what could be grown on neighboring hill land during the wet season, and what could be fished, hunted, and gathered, will have provided calorically rich staples, plant and animal protein, and a wide range of additional nutrients and amenities, as well as what was needed to pay the tribute.

"Basic agriculture" is a particularly useful synonym when one considers references to agricultural production after Contact. Sugarcane is an early preoccupation in Central Veracruz. Aside from that, a wide range of productive activities is implied in the observations of the authors of the *Declaración* and the *Relaciones*. These are considered Indian things and passed over lightly or taken for granted. It can hardly have been for want of possibilities all around to obtain basic necessities that these people diminished drastically.

Nineteenth-century observers are quite dismissive too. And yet, in their interstitial remarks they often backhandedly acknowledge basal agriculture. They see little sign of the plow and yet the markets are full of produce. Sartorius sketches the adroitness of the rural Veracruzanos well, in a mixture of scorn and envy. He sees great potential for various commercial crops but views as insignificant the agriculture that is actually going on.

The ingenieros of the latter half of the twentieth century were as frustrated by all the local ways of making-do as was Sartorius. Their tables of products are a cornucopia, but the ways in which all this is produced are labeled archaic. Production needs to be intensified and modernized.

Many aspects of the rural life sketched by Sartorius and deprecated by the ingenieros were found by the author in the Veracruzan backcountry of the 1960s: the refreshments offered on arrival at a rural dwelling, the chickens and pigs around the door, the curtains of dried beef overhead, the structure and the furnishings of the dwelling noticed during the rest and the chat with the hosts, the subsequent amble through gardens and what seemed scruffy little fields, filled with a jumble of all sorts of different plants. The hosts themselves were quite likely to apologize for all the backwardness, and the visitor, in return, would feel compelled to complement the productivity profusely. All this I can interpret now as representing an imposing complex of continuous sustaining activities similar to the historic and the prehistoric.

It has been pointed out that the "traditional" should not be sought in areas of recent in-migration, such as the immediate surroundings of the San Juan Basin (e.g., González Jácome n.d.: 3). This is reasonable, but perhaps not so compelling as it seems at first. In-migrants are likely to arrive without the locally traditional repertoire of skills, of course, but they may well bring with them the work and make-do inclinations of an agricultural household on a small holding. In any case, through learning, adaptation, or innovation, a functionally similar overall subsistence strategy can soon emerge. In addition, of course, everyone seeks for something commercial, something to finance manufactured necessities and overall economic improvement of the household. Cues are taken from those who have lived in the area longer, from what was known in the former home areas, perhaps even from an agrónomo passing by, or problems are solved by adaptation and innovation.

I observed such reconstitution in the colonies of in-migrants in Southern Veracruz during the 1960s. Their communities were rather miserable at first, and many people abandoned their new holdings. Others remained and adjusted, learning how to subsist, which sometimes meant "reverting" to shifting cultivation, and perhaps even to prosper, which usually meant getting into ranching. In the colonies along the Candelaria River in Campeche at about the same time, people could be seen here and there taking positive action, in one notable case cutting transportation canals across wetland to get to agriculturally promising hill land. They did not know that they had cut the canals where the ancients had done it before them, as could be seen from the air.

Paul Richards's excellent *Indigenous Agricultural Revolution* (1985) is relevant here. The book is based on African examples but the ideas have broader application. He shows that in many areas large-scale mechanized agriculture has had a poor record and that in such areas it is important to credit the potential for inventive self-reliance on small and, one might add, "traditional" holdings. Their substantial potential for development, not grand extraneous models, should be backed with foreign aid.

William Denevan, a highly respected student of prehistoric agricultural methods and mentor of all of those who have worked on "raised fields," notes that continuation of the prehistoric in the "traditional"—a term he is prepared to use without quotation marks—is a good indicator of sustainability (1995). The indications that wetland planting platforms in the Mexican lowlands lapsed long before conquest and were not revived later puts that form of wetland agriculture into a negative light, and that deserves further reflection. The related *chinampas*, of course, persisted.

In recent decades "traditional" ways in Mexico have become seriously constrained and indeed are officially scheduled now to be overridden by globalization. We are concerned over their passing, but perhaps we have underestimated the continuing need for them and their resiliency (Siemens 1996).

Representative Refractions

The investigation of the San Juan Basin began fairly empirically and was continued mostly in that vein, but representational issues often emerged and sometimes obtruded. It became progressively clearer as one worked to trace actual land-use continuities, discontinuities, and the processes of change that there was a succession of rather distinct renditions. Various "worlds" had been "written" for us—the reference is to *Writing Worlds*, a volume of essays edited and well introduced by Trevor Barnes and James Duncan (1992).

It is not difficult to see a series of "texts": the various Basins drawn from the documentary sources, but also what was on the air photographs and, indeed, what we ourselves deduced from the stratigraphic evidence and saw around us in the field. It is useful to follow Barnes and Duncan here verbatim:

> The notion of text used in this volume is not the traditional one of a printed page or a volume sitting on a shelf in the library. Rather, following Roland Barthes and other literary theorists and cultural anthropologists, we use an expanded concept of the text: one that includes other cultural productions such as paintings, maps and landscapes, as well as social, economic and political institutions. These should all be seen as signifying practises that are read, not passively, but as it were, rewritten as they are read. [They are] . . . constitutive of reality rather than mimicking it—in other words, as cultural practise of signification rather than as referential duplications . . . such practises

of signification are intertextual in that they embody other cultural texts and, as a consequence, are communicative and productive of meaning. Such meaning, however, is by no means fixed; rather, it is culturally and historically, and sometimes even individually and momentarily, variable. (Barnes and Duncan 1992: 5–6)

Given the postmodern winds felt everywhere, it has been difficult to avoid probing the textual strategies of the various "authors," including this one, in order to appreciate their renditions *per se*, to note the effects of the various kinds of evidence, to explore their complementarity and attempt a measure of reconciliation. The yields in this respect are more elaborative than strictly cumulative or directional, and they are quite inconclusive, but they are intriguing. As one turns the crystal, refractions shift, colors appear and disappear.

Broaching Lacunae

Perceptive separations have constrained our investigation of wetlands from the beginning. In the early forays within the Maya region, when we approached wetlands not yet cleared for ranching we would ask lowlanders about the patterning. They would be quite mystified; in their gestures and the way in which they talked about their occasional hunting and fishing trips into the wetlands they indicated that this was terrain where few people went. They would take us into various wetlands by canoe, threading waterways that they regarded as natural but which we might recognize as the remains of canals. Among the hydrophytes, often well above our heads, we were not many metres from firm ground but felt ourselves far removed. It was amazing to remember and difficult to explain to our hosts that these had once been gardens.

Along the margins of the floodplain of the San Juan River and many other similar places the landscape itself symbolizes and, indeed, enforces this distance between wetlands and the cultivated and occupied firm ground with a curious vegetational barrier (Figures 2.6, 2.14). A band of trees and tangled underbrush often makes it difficult to approach the wetlands, should one want to.

Actually wading into a swamp is not easy. It calls up a heritage of fear, embedded in our languages. In the early stages of our investigations in Northern Belize, we needed to move from the main streams into the forest in regularly spaced arbitrary directions in order to sense with our feet whatever remains of platforms and canals there might be. This would inevitably lead over levees into backswamps. We clutched our compasses and readily answered the shouts of unseen companions transecting along parallel lines on either side. Returning from such fieldwork, dirty and wet, we would soon transform the experience. By the time we had convened for happy hour at project headquarters the incursions had become the subject of competitive banter; willfully overcoming inherited aversions had become heroic.

These and subsequent attempts to take a swamp-centric view have had some good results, but they remain a bit of an affectation. Field slogging, careful ethnography, and textual analysis can mitigate or qualify, but not fully vitiate, that culture-historical "otherness."

Those who "write" the Central Veracruzan lowlands for us often miss or evade the bottomlands, regarding them, if they take note of them at all, from the outside. For the Europeans who approached the Mesoamerican coast, from the time of the earliest Spanish landfalls to the onrush of nineteenth-century travelers, the main point was usually to get up into the central highlands. You crossed the lowlands as expeditiously as possible and rounded the wetlands en route. They were a source of menace.

The vicar, the médico, and those who drew the forceful early maps are looking southward from old Veracruz. Medellín, their rival town, and San Juan de Ulúa, the main anchorage for trans-Atlantic shipping, lie to the south. The sluggish little river that drains the intervening land empties into the larger through-stream right there in front of the town. We are repeatedly taken into the old town of Veracruz, and the authors can hardly avoid taking some notice of the Basin, but its bottomlands remain distant in their prose, vague and generalized on their maps, or blank.

The perspective on the Basin changes, to the extent that anyone takes any notice, as the functions of old Veracruz shift to the new site opposite San Juan de Ulúa. Old Veracruz becomes sad and boring—a fishing village. Traffic out of the new Veracruz passes the Basin by to the south. Jalapa, the comfortable mid-slope site of the great fair, and Mexico City are the attractions, as before. There is no port or rival town to the north nor any other reason, really, to look in that direction. In the late-eighteenth century Corral views Central Veracruz from the coast. He maps it quite sharply but still describes it from the outside.

With the coming of independence one might have expected a swath of terrain along the road into the interior to resolve more clearly. The country had become easily accessible, visitors streamed in, and descriptions multiplied. However, they have little to offer specifically on the Basin and on the neighboring hill land, except one.

Sartorius finally gives the lacunae some detail; he lived upslope but was not afraid to travel through the San Juan Basin and he wrote about it with considerable competence and style. The fact that there are no maps is bothersome, but one can understand that Sartorius would not have the means with which to try to do anything like what Humboldt and his draftsmen had already done. Rugendas's images decorate the text and raise issues of their own but do not really illustrate it.

The ingenieros provide us with magnificent photographic and topographic raw materials, which were our guides for dry-season ground reconnaissance in the wetlands and basic to almost everything else we did as well. These authors do characterize the Basin in their natural environmental surveys, particularly in their observations on soils. One can tell, as they protest repeatedly, that they were in the field. But they seem to have hurried in and out of the bottomlands; they recommend drainage—they do not appreciate their significance, *per se*, for ranching or agriculture. The technicians certainly had little sense of history and cannot really have descended into their own air photographs and maps. For us there is a substantial discrepancy between what was and what might have been "read" out of the visual materials. They are a treasure and our fingers curl at the thought of their potential. In various fields to this day, they are seen as an optional extra, as decoration.

Considering the avoidance and aversion through historic times, the great need to drain, over against the evidence for intensive Prehispanic agriculture in the wetlands, one wonders about the attitudes of those who dug the canals and built the platforms. Surely what has been distant and sinister was once familiar, central, and sustaining. Was what historically has been repulsive once attractive? Did wetland rigors need to be overcome heroically in face of need? The late Alfonso Medellín Zenil, a thoughtful, sensitive scholar of Veracruzan prehistory, presents a tantalizing indication out of the Classic period at the site of Las Higueras, near the coast, on the very northern extremity of what may be regarded as Central Veracruz. He interprets a striking mural: it shows a god of inundation pouring water from a large jug on to the earth. Two fish are ascending the stream of water, and nearby stands a flourishing maize plant (1979: 211–213). Is this an image of domestication and plenty overshadowed by beneficence? Perhaps.

If the wetland incursions of the late Classic were responses to a concatenation of social and environmental stresses—a period of drying and reduced yields on the soils of the hill lands—and if the inhabitants of the lowlands at the time had similar apprehensions about wetlands as those who have lived around them in historic times, then the incursions themselves become an indicator of the seriousness of the pressure.

In any case, the considerable recent research into Prehispanic wetland agriculture has effected some reversal of the old disregard. The wetlands of the tropical lowlands have ceased to be just lacunae and become positive spaces, so much so, in fact, that exaggerated and oversimplified statements about the significance of the agriculture practiced within them must be qualified by some of us who participated in the early phases of this line of investigation.

Shifts in Terms of Reference

The Basic Scientific Study.   The research that underlies chapter 2, and perhaps other parts of this analysis, too, has been scientific in a rather unselfconscious way. I have taken some pains to learn about and to enlist collaboration in new techniques of diagnosis and measurement, which have repeatedly facilitated new insights on old and difficult issues. Care has been taken with the treatment of evidence, with deduction and induction, but one remains an amateur, hopefully in the best sense of the term.

There has been some jostling over evidence in symposia and in the literature. Direct evidence, like radiometrically dated remains of cultigens, while critically important, can be given inordinate importance over against context and analogues. Certain quantitative exercises regarding the extent and volume of planting platforms and canals have repeatedly proved irresistible but less than enlightening. Unfortunately erosion and deposition have smoothed the microtopography. Outlines of platform remains as reflected in the vegetation are thus not very reliable indications of previous extent—canal midlines are more reliable with respect to certain aspects of that system. How much was produced or how many people were sustained is just not reliably deduced. How much labor was required to build such systems is hard to determine when it is not known how the work was organized. Assumptions have to be piled on assumptions, diluting the meaning of any figures that may result. On the other hand, faces in the fraternity pale at the prospect of moving from the morphological to the cosmographic. In informal discussions, two veterans of archaeoastronomy, Avenyi and Tichy, have remarked on this curious phenomenon. In spite of clear indications of ancient means of integrating landscapes with the cosmos, and of the pervasiveness of the spiritual in prehistory, the tendency has usually been to remain materialistic.

Our analysis is not the categorical science of the ingenieros. There are few illusions about objectivity; the reasoning is not particularly positivistic. It is accepted that inquiries in the natural sciences involve questions of meaning, interpretation, and rhetoric (Johnston et al. 1994: 457). One is conscious of some of the recent vilification of science but not overwhelmed by it—that way lies paralysis.

From Prehistory into History.   Having decided to proceed diachronically, the investigator emerges from the ancient landscape into a place that is both younger and older. What seemed familiar becomes strange again, antique.

The careful handling and rather determined deciphering that the early colonial manuscripts require induce respect. Such solemn introductory invocations and self-possessed concluding signatures! When one has penetrated the legalistic and procedural matrix and comes to some actual description, it seems to have weight. In much of the published discussion of colonial sources, hard-won documentary information is presented as treasure, certainly in the very useful *Guide to Ethnohistorical Sources* in the *Handbook of Middle American Indians* (Cline 1972). It is difficult to resist this attitude; these are yields indeed.

The categories in the documents, however, are often uncongenial. I have been encouraged by some of the respondents to the set of questions that came from the Council of the Indies. This information that you require here, one of them will say, has already been provided under such and such; or, we will get to that other topic in due course—they were protesting against some of the nonsequiturs that had puzzled me!

So one reformats. A biological procedure initially suggested itself: going from common names of plants and animals to Latin designations and thence to ecology, but the gap between the common and the Latin proved too large for this study. Altitudinal schemata, however, could be used to link the many specific facts provided regarding natural plant communities, human activity, and material culture. Old maps and the written documents' own spatial indications helped, but it is often necessary to shift toponymy to a modern base. Sluyter has now done this better than anyone else (1995: 419–423).

From the Written to the Printed. Over against the treasured manuscripts, there are the relatively easily accessible nineteenth-century materials, often with self-effacing prefaces and a daily-journal format. They can easily seem ephemeral. They have been frequently anthologized and expected somehow, impressionistically, to add up. And, if not that, then at least to entertain.

Such a comparison is easily upended. Many of the old documents, such as one or another of the *Relaciones Geográficas*, were hastily written and mostly submitted, one deduces, as first drafts—the vicar did insert some additional comments! The maps, particularly those that accompany litigation over land, were often very sketchy approximations. Biases, idiosyncrasy, and imagination often show through, thank God. And in the background are medieval mind-sets.

The nineteenth-century materials were at least edited to one degree or another. In most authors' backgrounds there was romanticism and at least a glimmer of the new interest in the natural; they are much closer to us in their thinking than the colonial observers. One is particularly thankful for those who could be called naturalists—more sensitive to nature than to people but also unafraid to explore in terrain that had been given a very bad reputation, curious, observant, adumbrating issues that would come to dominate our worldview.

One of them merits special attention in our analysis. He has his cluster of biases, particularly a strong predisposition toward colonization, which proved unrealistic. However, this leads to a calm, deliberate, and, indeed, sympathetic rendition of tropical lowlands, which suits our purposes. He finds lowlanders unsympathetic, mostly, except rancheros and vaqueros. Sartorius has few doubts about *them*; they are fine people. And there is nothing better than life on the land.

His text ramifies absorbingly under scrutiny, especially in the original. Humboldt is embellished indeed in this portrayal of Mexico; he is bested too, in various respects. Sartorius does not have the intellectual scope nor the data that Humboldt had, but he does look carefully at some Mexican terrain. He does more than declaim in dramatic terms about altitudinal zonation, which was Humboldt's large theme; he details it. It must be noted that the country is seen from his property, El Mirador, near Huatusco; many a generalization is exemplified from Central Veracruz. Someone seeking another Mexican region would be disappointed.

Into a Bureaucratic Midden.   Reading the ingenieros is not like reading Sartorius; their prose is fibrous and their science naive. They seem to have opted for modernization without sufficient reflection. In this they hold up a mirror to one's own early work in lowland Veracruz at about the same time. They also beg a contemporary reaction, an attempt at a more ecologically articulate assessment of the nature and use of the lowlands as well as a reappreciation of the "traditional."

The most uncongenial aspect of the reports is their focus on the ills of this landscape, while we were exploring the realization of its favor! The diagnosis is too direct and the prescription too confident. The discussions with Juan Pablo Martínez, who was one of the ingenieros, have confirmed that the whole approach of the ministry to the definition and solution of developmental problems was culturally and socially insensitive (1994).

Although the ingenieros see the Basin as a problem area, they do map out an interesting geography of favor, which is diagnostic and sets our own into relief. They see the best lands, or rather the least impeded lands, as just above the mean level of benign flooding. Their great objectives are to bring down that level, to broaden the belt of favor, and to enhance production in that belt by bringing in irrigation water. Our "best lands" are just below the mean level of benign flooding, an assessment long shared by ranchers.

Returning to the Premise of this "Text"

The positive bias has been obvious since the title page. These are not threatening tropical lowlands, nor necessarily threatened, nor exotic. They are not the potentially productive but still difficult and underused or overused lowlands, but actually productive terrain, historically and prehistorically. And the wetlands are not excepted in the discussion of the lowlands; they are made to epitomize the productivity.

The aura of productivity is sustained by the early Spanish materials: the explorers' and conquerors' accounts in the first place, then the early chronicles and the *Relaciones,* and whatever other documents offer clues to the nature and use of the wetlands. Repeatedly one comes on general references to high and most agreeable productivity in the regions within which the wetlands are found—these were places where one could obtain food for the continuation of the *entradas!*

Several strong images strengthen the impression. The médico presents the "well-watered" Basin to us with the sweep of an Old Testament hand—the promised land as shown to Lot and Abraham (Genesis 13: 10–18). As one reads the lists and the descriptions of what is to be had in the New World, its products assemble themselves into a massive still life. The hunting and the eating are good.

The chroniclers trace the theme back into prehistoric times. Durán heard an Aztec informant describe the rich capacity of their former coastal homeland (Melgarejo 1980: 120). Legend or otherwise, the description parallels and lends most agreeable imagery to our evaluation of the evidence for wetland agriculture and related production on neighboring microenvironments; we have alluded to it on the first page of chapter 2. That lowland place, the informant recalled, had been a region of abundance. There had been plenty of water; the environment had offered food resources of many kinds. Most importantly, at any given time there had been some maize ripe and ready to pick, some with new ears, some just growing up, and some newly sprouted. In such a place there could never be hunger.

This image and our own attribution of interdigitated productivity to wetlands and their surroundings in prehistoric times may be too favorable by half, but such are the current indications. This might well have to be qualified if wetland agriculture is eventually shown, say, to have been a desperate concomitant of decline.

Prehistoric land use has certainly been compared favorably in this analysis with modern agricultural land use, which may be producing well for export but often appears less impressive in its sustenance of the resident population, at least the substantial marginalized portion of it. The emphasis on productivity is in part a reaction to past evaluations of land use in Mexico's tropical lowlands, in particular critical extraneous views of what needed to be done and, even more, the indictments of the lowlanders (Siemens 1990).

Post-contact productivity is realized largely in the context of ranching. Cattle in tropical pastures have become an ominous image, with good reason. However, accepting the economic context, they do represent a large part of the productivity of tropical lowlands. The early colonial observers depict the pastures and herds of Central Veracruz in exuberant terms. On the eve of independence Corral reports that cattle haciendas flourish and other resources beckon but that it is also a difficult region in which to live. Humboldt sees great potential, as do the naturalists that follow—their appreciation of productivity is clear in spite of the indictments that they freely distribute. And considerable agricultural and ranching productivity is apparent also underneath the ingeniero's pathology of the Basin. It is clearly a durable theme.

Also, I suppose, it needs to be admitted in conclusion that a positive gist was stimulated by the means with which the investigation of wetland patterning in Mesoamerica began. Aerial views inveigled, as they usually do. The material remains themselves presented a fascinating puzzle. Ground reconnaissance and talks in the tropical shade with knowledgeable people were refreshing after periods of sedentary academic life. Not difficult, then, to sustain a positive mind set! In any case, this approach facilitated investigation; it allowed an expansive interpretation of ancient subsistence and of the wetlands themselves as sustaining areas.

References

Acuña, René. 1985. *Relaciones Geográficas del siglo XVI*, Vol. 5/2: *Tlaxcala*. Instituto de Investigaciones Antropológicas, Serie Antropológica 59. Mexico City: Universidad Nacional Autónoma de México.

Adams, Richard E. W. 1977. *Prehistoric Mesoamerica*. Boston: Little, Brown.

———. 1980. "Ancient Land Use and Cultural History in the Pasion River Region." Reprint from Burg Wartenstein Symposium No 86 (August). New York: Wenner-Gren Foundation for Anthropology.

Adams, Richard E. W., W. E. Brown, Jr., and P. T. Culbert. 1981. "Radar Mapping, Archaeology and Ancient Maya Land Use." *Science* 213 (4515): 1457–1463.

Adams, Richard E. W., Patrick T. Culbert, Walter E. Brown, Jr., Peter D. Harrison, and Laura J. Levi. 1990. Commentary: "Rebuttal to Pope and Dahlin." *Journal of Field Archaeology* 17: 241–244.

Alemán Valdez, Miguel. 1986. *Remembranzas y testimonios*. Mexico, Barcelona, Buenos Aires: Grijalbo.

Del Angel-Pérez, Ana Lid. 1989. Personal communication.

———. 1991. *Manejo de recursos agro-ganaderos en la zona inundable de Veracruz Central*. Thesis, Universidad Iberoamericana, Mexico City.

———. 1991. Personal communication.

———. 1992. "Manejo de recursos y gandería en un ejido de Veracruz Central México." Unpublished manuscript in personal files of Alfred H. Siemens.

———. 1994. "Formación de la estructura productiva ganadera en la llanura costera de Veracruz central." In *Las llanuras costeras de Veracruz*. Edited by Odile Hoffmann and Emilia Velázquez. Xalapa, Veracruz: ORSTOM and Universidad Veracruzana.

Appenzeller, Tim. 1994. "Clashing Maya Superpowers Emerge from a New Analysis." *Science* 266(4):733–734.

Archivo General de Indias (AGI). 1777 [1864]. Foja 11v–57v. Mexico.

Archivo General de la Nación (AGN). 1979. *Catálogo de Ilustraciones*. Mexico City: Archivo General de la Nación.

Ariel de Vidas, Anath. 1994. "La bella durmiente: el norte de Veracruz." In *Las llanuras costeras de Veracruz*. Edited by Odile Hoffmann and

Emilia Velázquez. Xalapa, Veracruz: ORSTOM and Universidad Veracruzana.

Aveni, Anthony F. 1980. *The Skywatchers of Ancient Mexico*. Austin: University of Texas Press.

Aveni, Anthony F., and S. L. Gibbs. 1976. "On the Orientation of Pre-Columbian Buildings in Central Mexico." *American Antiquity* 41 (4): 510.

Barbosa-Ramírez, René. 1993. *La Ganadería privada y ejidal. Un estudio en Tabasco*. Chapingo, Mexico: Centro de Investigaciones Agrarias.

Barkin, David. 1991. *Un desarrollo distorsionado: la integración de Mexico a la economía mundial*. Mexico City: Siglo Veintiuno Editores.

Barnes, Trevor J., and James S. Duncan, eds. 1992. *Writing Worlds*. London: Routledge.

Beck, Hanno. 1985. "Alexander von Humboldts Beitrag zur Kartographie." In *Alexander von Humboldt, Leben und Werk*. Edited by Wolfgang-Hagen Hein. Frankfurt am Main: Weisbecker Verlag.

Beltrán, Gonzalo Aguirre. 1991. *Regiones de Refugio*. Mexico City: Fondo de Cultura Económica.

Bentham, Jeremy. 1962. *Works*. New York: Russell and Russell.

Brading, David, ed. 1985. *Caudillos y campesinos en la Revolución Mexicana*. Mexico City: Fondo de Cultura Económica.

Butzer, Karl W. 1988. "Cattle and Sheep from Old to New Spain: Historical Antecedents." *Annals of the Association of American Geographers* 78(1): 29–56.

———. 1992. "From Columbus to Acosta: Science, Geography, and the New World." In *The Americas Before and After 1492: Current Geographical Research*. Edited by Carl W. Butzer. *Annals of the Association of American Geographers* 82(3). Washington, D.C.: Association of American Geographers.

Butzer, Karl W., and Elizabeth K. Butzer. 1995. "Transfer of Mediterranean Livestock Economy to New Spain: Adaptation and Consequences." In *Global Land Use Changes: A Perspective from the Columbian Encounter*. Edited by B. L. Turner II, Antonio Gómez Sal, Fernando González Bernáldez, and Francisco di Castri. Madrid: Consejo Superior de Investigaciones Científicas.

Byrne, Roger. 1992. Personal communication.

Byrne, Roger, and Sally P. Horn. 1989. "Prehistoric Agriculture and Forest Clearance in the Sierra de los Tuxtlas, Veracruz, Mexico." *Palynology* 13: 181–193.

Calcott, Wilfrid Hardy. 1936. *Santa Anna*. Norman: University of Oklahoma Press.

Cambrezy, Luc, and Bernal Lascuráin. 1992. *Crónicas de un territorio fraccionado: de la hacienda al ejido (Centro de Veracruz)*. Mexico: Larousse.

Carlson, John B. 1976. "Astronomical Investigations and Site Orientation Influences at Palenque." In *The Art, Iconography, and Dynastic History of Palenque*, Part III: *Proceedings of the Segunda Mesa Redonda de*

Palenque, Dec. 14–21. Edited by Merle Greene Robertson. Pebble Beach, Calif.: Robert Louis Stevenson School.

———. 1977. "The Case for Geomagnetic Alignments of Precolumbian Mesoamerican Sites." *Kathrob—A Newsletter-Bulletin on Mesoamerican Anthropology* 10, 2 (June): 67–88.

———. 1981. "A Geomantic Model for the Interpretation of Mesoamerican Sites: An Essay in Cross-Cultural Comparison." In *Proceedings of the Dumbarton Oaks Conference on Mesoamerican Sites and World-Views 1976.* Washington, D.C.: Trustees for Harvard University.

Carroll, Patrick J. 1991. *Blacks in Colonial Veracruz: Race, Ethnicity, and Regional Development.* Austin: University of Texas Press.

Chevalier, François. 1963. *Land and Society in Colonial Mexico: The Great Hacienda.* Berkeley: University of California Press.

———.1970. *Land and Society in Colonial Mexico.* Berkeley: University of California Press.

Chevey, P., and F. Le Poulain. 1940. "La Peche Dans les Eaux Douces du Cambodge." *Memoire de L'Institut Oceanographique de L'Indochine* 5:193.

Cisneros Solano, Victor M., Dámaso Martínez Pérez, Salvador Días Cárdenas, José Antonio Torres Rivers, Carlos Guadarrama Zugasti, and Artemio Cruz León. 1993. *Caracterización de la agricultura de la zona central de Veracruz.* Universidad Autónoma Chapingo, Mexico.

Cline, Howard F. 1961. "The Patiño Maps of 1580 and Related Documents; Analysis of 16th Century Cartographic Sources for the Gulf Coast of Mexico." *México Antiguo* 9:633–687.

———. 1972. "The Relaciones Geográficas of the Spanish Indies, 1577–1648." In *Handbook of Middle American Indians,* Vol. 12, Part 1: *Ethnohistorical Sources.* Edited by Howard F. Cline. (Including, as Appendix B: "Instruction and memorandum for preparing the reports which are to be made for the description of the Indies that his majesty commands to be made, for their good government and ennoblement.") Austin: University of Texas Press.

Clutton-Brock, Juliet. 1981. *Domesticated Animals from Early Times.* London: British Museum (Natural History).

Coe, Michael D., and Richard A. Diehl. 1980. *In the Land of the Olmec,* Vol. 2: *The People of the River.* Austin: University of Texas Press.

Compañía Mexicana Aerofoto (CMA). 1973. La Antigua/Actópan series of air photographs, 1:8000, and Levantamiento Topográfico, Distritos de Riego Actópan y La Antigua, Veracruz, 1:5000. Photographs made for Secretaría de Recursos Hidráulicos.

Corral, Miguel del. 1777. *Relación.* Archivo General de Indias, México, Legajo 1864, f. 12–57.

———. 1771. *Relación.* Archivo General de la Nación, Historia, Vol. 359, Exp. 1, 2. Fojas 1–17.

Cortés, Hernan. 1972. *Letters from Mexico.* Translated by A. R. Pagden. London: Oxford University.

Dalcq, Albert M. 1951. "Form, and Modern Embryology." In *Aspects of Form*. Edited by Lancelot Law Whyte. Lund: Humphries.

Dampier, William. 1906. *Dampier's Voyages*. Edited by John Masefield. New York: E. P. Dutton.

Darch, Janice P. 1980. "Soils. The Maya Raised-field Agriculture and Settlement at Pulltrouser Swamp, Northern Belize." Unpublished report of Pulltrouser Swamp Project submitted by University of Oklahoma, Norman, to National Science Foundation.

——, ed. 1983. *Drained Field Agriculture in Central and South America*. BAR International Series 189. Oxford: BAR.

Dary, David. 1981. *Cowboy Culture*. New York: Knopf.

De la Mota y Escobar, Fray Alonso. 1987. "Memoriales del obispo de Tlaxcala: Un recorrido por el centro de México a principios del siglo XVII." In *Memoriales del obispo de Tlaxcala*. Edited by Alba González Jácome. Mexico City: Secretaría de Educación Pública.

De la Torre, Mario, ed. 1985. *El México Luminoso de Rugendas*. Mexico City: Cartón y Papel de México.

Décamps, Henri, and Robert J. Naiman. 1990. "Towards an Ecotone Perspective." In *The Ecology and Management of Aquatic-Terrestrial Ecotones*. Edited by Robert J. Naiman and Henri Décamps. Paris: UNESCO and Parthenon Publishing Group.

Devane, William. 1995. "Prehistoric Agricultural Methods as Models for Sustainability." *Advances in Plant Pathology* 11: 21–43.

Díaz del Castillo, Bernal. 1963. *The Conquest of New Spain*. Harmondsworth, Middlesex: Penguin.

Dobie, J. Frank. 1980. *The Longhorns*. Austin: University of Texas Press.

Domenech, Manuel. 1922. *México tal cual es: 1866*. Querétaro: Demetrio Contreras.

Doolittle, William E. 1987. "Las Marismas to Pánuco to Texas: The Transfer of Open Range Cattle Ranching from Iberia through Northeastern Mexico." *Conference of Latin Americanist Geographers Yearbook* 13: 3–11.

Doolittle, William E., and Charles D. Fredrick. 1991. "Phytoliths as Indicators of Prehistoric (*Zea mays* subsp. *mays*, Poaceae) Cultivation." *Plant Systematics and Evolution* 177: 175–184.

Eckholm, G. F. 1944. *Excavations at Tampico and Panuco in the Huasteca, Mexico*. American Museum of Natural History, Anthropological Papers 38.

Eisely, Edwin. 1961. *Darwin's Century*. Garden City, New York: Doubleday/Anchor.

Erickson, Clark L. 1985. "Applications of Prehistoric Andean Technology." In *Prehistoric Intensive Agriculture in the Tropics*. Edited by I. S. Farrington. BAR International Series 232(I): 209–232.

Fedick, Scott L. 1995. "Indigenous Agriculture in the Americas." *Journal of Archaeological Research* 3(4): 257–303.

Foote, Kenneth E. 1985. "Space, Territory, and Landscape: The Borderlands of Geography and Semiotics." *Recherches Semiotiques/Semiotic Inquiry* 5 (2): 158–175.

Foucault, Michele. 1979. *Discipline and Punish: The Birth of the Prison.* New York: Vintage.

Fowler Salamini, Heather. 1979. *Movilización campesina en Veracruz, 1920–1938.* Mexico City: Siglo Veintiuno Editores.

Fry, Robert E. 1983. "The Ceramics of the Pulltrouser Area: Settlements and Fields." In *Pulltrouser Swamp: Ancient Maya Habitat, Agriculture, and Settlement in Northern Belize.* Edited by B. L. Turner II and Peter D. Harrison. Austin: University of Texas Press.

García, Enriqueta. 1976. "Los Climas del Estado de Veracruz." *Anuario del Instituto Biológico de la Universidad Nacional Autónoma, México: Seria Botánica* 41: 3–42.

Gemelli Carreri, Juan Francisco. 1983. *Viaje por la nueva España.* Mexico City: Jorge Porrua.

Gerhard, Peter. 1972. *A Guide to the Historical Geography of New Spain.* Cambridge University Press, Cambridge.

———. 1986. *Geografía Histórico de la Nueva España, 1519–1821.* Mexico City: Universidad Nacional Autónoma de México.

Glacken, Clarence J. 1967. *Traces on the Rhodian Shore.* Berkeley: University of California Press.

Gliessman, S. R., R. García E., and M. Amador A. 1981. "The Ecological Basis for the Application of Traditional Agricultural Technology in the Management of Tropical Agro-ecosystems." *Agro-Ecosystems* 7: 173–185.

Gliessman, Stephen R., B. L. Turner II, F. J. Rosado May, and M. F. Amador. 1983. "Ancient Raised Field Agriculture in the Maya Lowlands of Southeastern Mexico." In *Drained Field Agriculture in Central and South America.* Edited by J. P. Darch. BAR International Series 189. Oxford: BAR.

Goman, Michelle. 1992. *Paleoecological Evidence for Prehistoric Agriculture and Tropical Forest Clearance in the Sierra de los Tuxtlas, Veracruz, Mexico.* Master's thesis, University of California, Berkeley.

Gómez-Pompa, Arturo. 1973. "Ecology of the Vegetation of Veracruz." In *Vegetation and Vegetational History of Northern Latin America.* Edited by A. Graham. Amsterdam: Elsevier.

———. 1977. *Ecología de la Vegetación del Estado de Veracruz.* Mexico City: Instituto de Investigaciones sobre Recursos Bióticos.

———. 1980. *Ecología de la Vegetación del Estado de Veracruz.* Mexico City: Compañia Editorial Continental.

González Jácome, Alba. 1988. *Población, ambiente y economía en Veracruz central durante la colonia.* Mexico City: Universidad Iberoamericana, Fondo de las Naciones Unidas para actividades en material de población.

———. 1993. "Management of Land, Water and Vegetation in Traditional Agro-ecosystems in Central Mexico." *Landscape and Urban Planning* 27: 141–150.

————. 1997. "Algunas cuestiones sobre el ambiente, la población y la economía en Veracruz Central." In *Agricultura y Sociedad en México: Diversidad, Enfoques, Estudios de Caso.* Silvia del Amo Rodriguez (ed.) Plaza y Valdéz, México.

————. n.d. *Ambiente y Caña de Azúcar en Veracruz. Una Historia de la Agricultura Comercial y el Caso del.* Serie Antropología, Departamento de Ciencias Sociales y Políticas, Universidad Iberoamericana. Mexico City.

Gourou, Pierre. 1966. *The Tropical World: Its Social and Economic Conditions and Its Future Status.* New York: Wiley.

Greenblatt, Stephen. 1991. *Marvelous Possessions: The Wonder of the New World.* Chicago: University of Chicago Press.

Guevara, S., J. Meave, P. Moreno-Casasola, and J. Baborde. 1992. "Floristic Composition and Structure of Vegetation under Isolated Trees in Neotropical Pastures." *Journal of Vegetation Science* 3:655–664.

Gunn, Joel D., William J. Folan, and Hubert R. Robichaux. 1995. "A Landscape Analysis of the Candelaria Watershed in Mexico: Insights into Paleoclimates Affecting Upland Horticulture in the Southern Yucatan Peninsula Semi-Karst." *Geoarchaeology* 10 (1): 3–42.

Gutiérrez, Celso, and M. G. Zolá. 1987. "Hidrófitas de Nevería, Veracruz, México." *Biótica* 12: 21–33.

Hakluyt, Richard. 1927. *The Principal Navigations, Voyages, Traffiques, and Discoveries of the English Nation.* New York: Dent.

Hammond, Norman. 1980. "Prehistoric Human Utilization of the Savanna Environments of Middle and South America." In *Human Ecology in Savanna Environments.* Edited by David R. Harris. London: Academic Press.

Hammond, R., and P. S. McCullagh. 1974. *Quantitative Techniques in Geography.* Oxford: Clarendon Press.

Harley, J. Brian. 1989. "Deconstructing the Map." *Cartographica* 26 (2): 1–20.

————. 1992. "Rereading the Maps of the Columbian Encounter: The Americas before and after 1492." Current Geographical Research. *Annals of the Association of American Geographers* 82(3): 522–536.

Harris, David R. 1980. "Tropical Savanna Environments: Definition, Distribution, Diversity, and Development." In *Human Ecology in Savanna Environments.* Edited by David R. Harris. London: Academic Press.

Harrison, Peter D. 1978. "*Bajos* Revisited: Visual Evidence for One System of Agriculture." In *Prehispanic Maya Agriculture.* Edited by Peter D. Harrison and B. L. Turner II. Albuquerque: University of New Mexico Press.

Hebda, Richard J., Alfred H. Siemens, and Alastair Robertson. 1991. "Stratigraphy, Depositional Environment, and Cultural Significance of Holocene Sediments in Patterned Wetlands of Central Veracruz, Mexico." *Geoarchaeology* 6(1): 16–84.

Henderson, John S. 1979. *Atopula, Guerrero, and Olmec Horizons in Meso-

america. New Haven: Department of Anthropology, Yale University.

Hernández, Arias. 1571. "Declaración sobre Veracruz/Apuntes para la Descripción de Veracruz." In *Papeles de Nueva España*, 1905/1947. Edited by Francisco Paso y Troncoso. Mexico City: Biblioteca Aportación Histórica.

Hernández Diosdado, Alonso. 1580. "Relación de la Ciudad de la Veracruz y su Comarca." In *Relaciones Geográficas del siglo XVI*, Vol. 5/2: *Tlaxcala*. Edited by René Acuña. 1985. Instituto de Investigaciones Antropológicas, Seria Antropológica 59. Mexico City: Universidad Nacional Autónoma de México, 1985.

Herrera y Lasso, José. 1994. *Apuntes sobre irrigación. Notas sobre su organización económica en el extranjero y en el país*. Instituto Mexicano de Technología del Agua and Centro de Investigaciones y Estudios Superiores en Antropología Social, 1919. Tlalpan D.F., Mexico.

Hippocrates. 1939. *The Genuine Works of Hippocrates*. Translated by Francis Adams. Baltimore: Williams and Wilkins.

Hiraoka, Mário. 1985. "Mestizo Subsistence in Riparian Amazonia." *National Geographic Research* 1(2): 236–246.

Hodder, Ian. 1986. *Reading the Past*. Cambridge: Cambridge University Press.

Hodell, David A., Jason H. Curtis, and Mark Brenner. 1995. "Possible Role of Climate in the Collapse of Classic Maya Civilization." *Nature* 375 (June): 391–394.

Hoffmann, Odile, and Emilia Velázquez, eds. 1994. *Las llanuras costeras de Veracruz*. Xalapa, Veracruz: ORSTOM.

Holland, M. M., and J. Balco. 1985. "Management of Fresh Waters: Input of Scientific Data into Policy Formulation in the United States." *Verhandlungen Internationale Vereinigung Limnologie* 22: 2221–2222.

Homer. 1965. *The Odyssey*. New York: Airmont.

Von Humboldt, Alexander. 1969. *Mexico-Atlas*. Stuttgart: Brockhaus.

———. 1911. *Political Essay on the Kingdom of New Spain*. Book 3. Edited by J. Black. London: Longman.

Hurault, J. 1965. "Les principaux types de peuplement du sud-est Dahomey; et leur representation cartographique." [Paris, Inst. Geogr. Nat.] *Etud. Photo-interpret* 2: 79.

Iglesias, José María. 1966. *Acayucán en 1831*. Colección Suma Veracruzana. Mexico City: Editorial Citlaltépetl.

Instituto Nacional de Estadística Geográfica e Informatica (INEGI). 1981. *Carta Topográfica 1:250,000; Veracruz E14–3*.

———. 1987. *Carta Geológica 1:250,000; Veracruz E14–3*.

Jacob, J. S., and C. T. Hallmark. 1996. "Holocene Stratigraphy of Cobweb Swamp, a Maya Wetland in Northern Belize." *Geological Society of America Bulletin* 108 (7): 883–891.

Jamieson, Milton. 1849. *Journal and Notes of a Campaign in Mexico*. Cincinnati: Ben Franklin Printing House.

Jiménez-Osornio, Juan J., and Arturo Gómez-Pompa. 1987. "Las Chinampas Mexicanas." Paper presented at conference, Ecología y Recursos Naturales: Distrucción, Recuperación y Desarrollo en Iberamérica, Seville, Oct. 5–9.

Johnston, R. J., Derek Gregory and David Smith, eds. 1994. *The Dictionary of Human Geography*. Oxford: Blackwell.

Jordan, Terry G. 1989. "An Iberian Lowland/Highland Model for Latin American Cattle Ranching." *Journal of Historical Geography* 15 (2): 111–125.

————. 1993. *North American Cattle-Ranching Frontiers: Origins, Diffusion, and Differentiation*. Albuquerque: University of New Mexico Press.

Junk, Wolfgang J., Peter B. Bayley, and Richard E. Sparks. 1989. "The Flood Pulse Concept in River-Floodplain Systems." In *Proceedings of the International Large River Symposium, Honey Harbour, Ontario, Canada. Sept. 14–21, 1986*. Edited by Douglas P. Dodge. Canadian Special Publication of Fisheries and Aquatic Sciences 106. Ottawa.

Keen, Benjamin. 1992. *A History of Latin America*. 4th ed. Boston: Houghton Mifflin.

Kelly, Isabel, and Angel Palerm. 1952. *The Tajin Totonac, Part 1: History, Subsistence, Shelter, and Technology*. Smithsonian Institution, Institute of Social Anthropology, Publication 13. Washington, D.C.

Kenly, John Reese. 1873. *Memoirs of a Maryland Volunteer: War with Mexico in the Years 1846–7–8*. Philadelphia: Lippincott.

Kolata, Alan, O. Rivera, and J. Guzmán. 1989. "Experimental Rehabilitation of Tiwanaku Raised Field Systems in the Lake Titicaca Basin of Bolivia." In *The Archaeology and Paleoecology of Lukurmata, Bolivia*. Edited by Alan Kolata. Second Preliminary Report of Proyecto Wila Jawira. Chicago: Department of Anthropology, University of Chicago.

Kozlowski, T. T. 1984. *Flooding and Plant Growth*. New York: Academic Press.

Lawrence, E. A. 1982. *Rodeo: An Anthropologist Looks at the Wild and the Tame*. Knoxville: University of Tennessee Press.

López de Gómara, Francisco. 1964. *Cortés*. Berkeley: University of California Press.

Lundell, Cyrus L. 1937. *The Vegetation of the Petén*. Carnegie Institution, Publication 478. Washington, D.C.

McKnight, Tom. 1976. *Friendly Vermin: A Survey of Feral Livestock in Australia*. University of California Publications in Geography 21. Berkeley: University of California Press.

Marcus, Joyce. 1983. "Lowland Maya Archaeology at the Crossroads." *American Antiquity* 48 (3): 454–488.

Maria Dorronsoro, José. 1964. "La mechanización de la agricultura en los distritos de riego en México." *Ingeniería Hidraulica en México* (Jan.–June): 102–111.

Markman, Peter T., and Roberta H. Markman. 1989. *Masks of the Spirit: Image and Metaphor in Mesoamerica*. Berkeley: University of California Press.

Martínez Davila, Juan Pablo. 1994. Personal communication.

"Mayans Used Canals to Avert Food Crisis: Radar." 1980. *Capital Times,* Washington, June 3, p. 5.

"The Mayas Secret: A Canal Network." 1980. *Newsweek,* June 6, p. 24.

Mayer, Brantz. 1844. *Mexico as It Was and as It Is.* London: Wiley and Putnam; New York: Winchester.

Medellín Zenil, Alfonso. 1979. "Clásico Tardio en el centro de Veracruz." *Cuadernos Antropológicos* 2: 205–213.

Melgarejo Vivanco, José Luís. 1979. *Los Jarochos.* Xalapa: Gobierno del Estado de Veracruz.

———. 1980. *Antigua Ecología indígena en Veracruz.* Xalapa: Gobierno del Estado de Veracruz.

———. 1981. *Historia de la Ganadería en México.* Xalapa: Ediciones del Gobierno de Veracruz.

Melville, Roberto. 1994. *El concepto de cuencas hidrográficas y la planificación del desarollo regional.* CIESAS/Jalapa, Veracruz, September 26–28. Unpublished manuscript in personal files of Alfred H. Siemens.

von Mentz, Brígida. 1982. *México en el siglo XIX visto por los alemanes.* Mexico City: Universidad Nacional Autónoma de México.

———. 1990. "Estudio preliminar." In *México hacia 1850.* Edited by Carl Christian Sartorius. Mexico City: Consejo Nacional para la Cultura y las Artes.

Meyer, Michael C., and William L. Sherman. 1979. *The Course of Mexican History.* New York: Oxford University Press.

Moliner, María. 1992. *Diccionario de Uso del Español.* Vol. H–Z. Madrid: Gredos.

von Müller, Baron J. W. 1864. *Reisen in den Vereinigten Staaten, Canada und Mexico.* Leipzig: Brockhaus.

Munton, P. N., J. Clutton-Brock, and M. R. Rudge. 1984. "Introduction to Workshop on Feral Mammals." In *Feral Mammals—Problems and Potential.* Proceedings of Workshop on Feral Mammals, III International Theriological Conference, Helsinki, August 1982. Marges: International Union for Conservation of Nature and Natural Resources.

Murra, John V. 1978. *La Organización Económica del Estado Inca.* Mexico City: Siglo Veintiuno Editores.

Navarrete Hernández, Mario. 1983. Personal communication.

Niederberger, Christine. 1979. "Early Sedentary Economy in the Basin of Mexico." *Science* 203: 131–142.

Ochoa, Lorenzo. 1994. "Agricultura intensiva en el área maya: algunos cuestionamientos." In *Agricultura indígena: pasado y presente.* Edited by Teresa Rojas Rabiela. Mexico City: Casa Chata.

———(ed.). 1989. *Huaxtecos y Totonacos.* Mexico City: Consejo Nacional para la Cultura y las Artes.

Odum, E. P. 1971. *Fundamentals of Ecology.* 2d ed. Philadelphia: Saunders.

Olguín Palacios, Carlos. 1991. Personal communication.

Orive Alba, Adolfo. 1947. "Programa de irrigación del C. Presidente

Miguel Alemán." *Ingeniería Hidráulica en México.* Febrero–marzo: 17–32.

———. 1990. Personal communication.

Orozco-Segovia, A., and S. Gliessman. 1979. "The Marceño in Flood-prone Regions of Tabasco, Mexico." Paper presented at Symposium on Mexican Agroecosystems, XLVIII Congress of Americanists, Vancouver, Aug. 11–17.

Ortega y Medina, Juan A. 1955. *México en la conciencia anglosajona.* México y lo Mexicano Series, no. 22. Mexico City: Robredo.

Oxford English Dictionary (Compact Edition). 1982. New York: Oxford University Press.

Palacios, Leopoldo. 1994. *El problema de la irrigación.* Jiutepec, Morelos: Instituto Mexicano de Tecnología del Agua; Tlalpan: Centro de Investigaciones y Estudios Superiores en Antropología Social.

Palerm, Angel, and Eric R. Wolf. 1957. "Ecological Potential and Cultural Development in Mesoamerica." In *Studies in Human Ecology.* Edited by L. Krader and Angel Palerm. Washington D.C.: Anthropological Society of Washington and General Secretariat of the Organization of American States.

Parsons, James J. 1989. *Hispanic Lands and Peoples: Selected Writings of James J. Parsons.* Edited by William M. Denevan. San Francisco: Westview.

Parsons, J. R., E. Brumfiel, M. H. Parsons, and D. J. Wilson. 1982. *Prehispanic Settlement Patterns in the Southern Valley of Mexico: The Chalco-Xochimilco Region.* Memoirs of the Museum of Anthropology, no. 14. Ann Arbor: University of Michigan.

Parsons, Talcott. 1937[1949]. *The Structure of Social Action: A Study in Social Theory with Reference to a Group of Recent European Writers.* New York: The Free Press of Glencoe.

Paso y Troncoso, Francisco del (ed.). 1947. *Papeles de Nueva España.* Mexico City: Biblioteca Aportación Histórica.

Pasquel, Leonardo (ed.). 1958. *La Ciudad de Veracruz.* 2 vols. Mexico City: Editorial Citlaltépetl.

———. 1963. *Miguel del Corral: La costa de sotavento.* Collección Veracruzana. Mexico City: Editorial Citlaltépetl.

———. 1979. *Viajeros en el Estado de Veracruz.* Mexico City: Editorial Citlaltépetl.

Petts, Geoff, and Ian Foster. 1985. *Rivers and Landscape.* London: Edward Arnold.

Pferdekampf, Wilhelm. 1958. *Auf Humboldts Spuren.* Munich: Max Hueber Verlag.

Piperno, Dolores R. 1988. *Phytolith Analysis: An Archaeological and Geological Perspective.* New York: Academic Press.

Pohl, Mary. 1985. "An Ethnohistorical Perspective on Ancient Maya Wetland Fields and Other Cultivation Systems in the Lowlands." In *Prehistoric Lowland Maya Environment and Subsistence Economy.* Edited by Mary Pohl. Cambridge: Harvard University Press.

————(ed.). 1985. *Prehistoric Lowland Maya Environment and Subsistence Economy.* Cambridge: Harvard University Press.

Pohl Deland, Mary. 1990. *Ancient Maya Wetland Agriculture: Excavations on Albion Island, Northern Belize.* Boulder: Westview Press.

Pope, Kevin O., and Bruce H. Dahlin. 1989. "Ancient Maya Wetland Agriculture: New Insights from Ecological and Remote Sensing Research." *Journal of Field Archaeology* 16: 87–106.

Pratt, Mary Louise. 1992. *Imperial Eyes: Travel Writing and Transculturation.* London and New York: Routledge.

Puleston, Dennis E. 1977. "Experiments in Prehistoric Raised Field Agriculture: Learning from the Past." *Journal of Belizean Affairs* 5: 36–43.

————. 1978. "Terracing, Raised Fields, and Tree Cropping in the Maya Lowlands: A New Perspective of the Geography of Power." In *Prehispanic Maya Agriculture.* Edited by Peter D. Harrison and B. L. Turner II. Albuquerque: University of New Mexico Press.

Raisz, Erwin. 1959. *Landforms of Mexico.* Prepared for Geography Branch of Office of Naval Research. Cambridge, Mass.: E. Raisz.

Ramírez Cabañez, Joaquín. 1943. *La Ciudad de Veracruz en el siglo XVI.* Mexico City: Imprenta Universitaria.

Rees, Peter. 1976. *Transportes y comercio entre Mexico y Veracruz, 1519–1910.* Mexico City: SepSetentas.

Religion and Agriculture in Mesoamerica. 1983. *New Scientist,* May 5, p. 291.

Richards, Paul. 1985. *Indigenous Agricultural Revolution.* Boulder: Westview Press.

Rios, Valencia. 1985. 3 Variedades de Cebú Revolucionaron la Industria Pecuaria de México. *El Dictamen,* June 8, pp. A1, A8.

Robertson, Alastair J. 1983. *Chinampa Agriculture: The Operation of an Intensive Pre-Industrial Resource System in the Valley of Mexico.* Master's thesis, Department of Geography, University of British Columbia, Vancouver.

Robertson, Donald. 1972. "The Pinturas (Maps) of the Relaciones Geográficas, With a Catalog." In *Handbook of Middle American Indians,* Vol. 12, Part 1: *Guide to Ethnohistorical Sources.* Edited by Howard F. Cline. Austin: University of Texas Press.

Roosevelt, Anna Curtenius. 1980. *Parmana: Prehistoric Maize and Manioc Subsistence along the Amazon and Orinoco.* New York: Academic Press.

Rzedowski, Jerzy. 1978. *Vegetación de Mexico.* Mexico City: Limusa.

Sanderson, Steven E. 1986. *The Transformation of Mexican Agriculture: International Structure and the Politics of Rural Change.* Princeton: Princeton University Press.

Santamaría, Francisco J. 1978. *Diccionario de Mejicanismos.* Mexico City: Editorial Porrua.

Santley, R. S. 1983. Personal communication.

Sargent, Frederick, II. 1982. *Hippocratic Heritage.* New York: Pergamon Press.

Sarmiento, G., and M. Monasterio. 1975. "A Critical Consideration of the Environmental Conditions Associated with the Occurrence of Savanna Ecosystems in Tropical America." In *Tropical Ecological Systems*. Edited by Frank B. Golley and Ernesto Medina. New York: Springer Verlag.

Sartorius, Carl. 1850. *Mexico als Ziel für deutsche Auswanderung*. Darmstadt: Reinhold von Auw.

———. 1859. *Mexiko. Landschaftsbilder und Skizzen aus dem Volksleben*. Darmstadt: Gustav Georg Lange.

———. 1870. "Memoria sobre el estado de la agricultura en el Partido de Huatusco." *Buletín de la Sociedad Mexicana* Vol. II: 141–197.

———. 1961. *Mexico about 1850*. Stuttgart: Brockhaus.

———. 1869. "Fortificaciones Antiguas." *Boletín de la Sociedad Mexicana*. 2d series, Vol. 1: 818–827.

———. 1990. *México hacia 1850*. Mexico City: Consejo Nacional para la Cultura y las Artes.

———. 1991. *México: Paisajes y Bosquejos Populares*. Mexico City: Centro de Estudios de História de México.

Scarborough, Vernon L. 1983. "A Preclassic Maya Water System." *American Antiquity* 48 (4): 720–744.

Scarborough, Vernon L., and Barry L. Isaac, eds. 1993. *Economic Aspects of Water Management in the Prehispanic New World*. Greenwich, Conn.: JAI Press.

Schmidt, Peter J. 1977. "Un sistema de cultivo intensivo en la cuenca del río Nautla, Veracruz." *Boletín del INAH* 20: 50–60.

Scholes, France V., and Ralph L. Roys. 1968. *The Maya Chontal Indians of Acalan-Tixchel*. Norman: University of Oklahoma Press.

Scholes, France V., and Dave Warren. 1968. "The Olmec Region at Spanish Contact." *Handbook of Middle American Indians*, Vol. 3, Part 2: *Archaeology of Southern Mesoamerica*. General Editor, Robert Wauchope. Austin: University of Texas Press.

Secretaría de Programmación y Presupuesto (INEGI). 1984. *Carta geológica*. I: 250,000. Veracruz E14-3. México.

Secretaría de Recursos Hidráulicos [SRH]. 1947. *Estudio Agrológico de Reconocimiento de la Margen Derecha del Río de La Antigua, Ver. Distrito de Riego de La Antigua, Ver*. Mexico City: Estudios y Projectos S.A.

———. 1971a. *Boletín Hidrológico* 38.

———. 1971b. *Boletín Hidrológico* 43.

———. 1973. *Estudio Socioeconomico del Proyecto de Rehabilitación de los distritos de riego "La Antigua" y "Actopan," Ver*. Mexico City: Estudios y Projectos S.A.

———. 1976a. *Estudio agrológico detallado distritos de riego no. 65 Actopan y 35 La Antigua, Ver. Memoria General*. Mexico City: Estudios y Projectos S.A.

———. 1976b. *Estudio agrológico detallado distritos de riego no. 65 Ac-*

topan y 35 La Antigua, Ver. Anexo no. 1. Mexico City: Estudios y Projectos S.A.

———. 1976c. "Plano Catastral 1970." In *Plan Nacional Hidráulico. Distritos de Riego "La Antigua y Actopan."* Mexico City: SRH.

———. 1976d. *La Obra Hidráulica de México a traves de los informes presidenciales.* Vols. 1 and 2. Mexico City. SRH.

———. n.d. *Estudio de Factibilidad del proyecto para la rehabilitación y ampliación de los distritos de riego la Antigua y Actopan.* (Undated, but apparently issued in the mid 1970s.) Mexico City: Estudios y Projectos S.A.

Sharrer, Beatriz. 1980. *La hacienda El Mirador, la historia de un emigrante alemán en el siglo XIX.* Thesis, Universidad Autónoma Metropolitana, Unidad Iztapalapa, Mexico.

Siemens, Alfred H. 1966. "New Agricultural Settlement along Mexico's Candelaria River." *Inter-American Economic Affairs* 20: 23–39.

———. 1978. "Karst and Prehispanic Maya in the Southern Lowlands." In *Prehispanic Maya Agriculture Albuquerque.* Edited by Peter D. Harrison and B. L. Turner II. Albuquerque: University of New Mexico Press.

———. 1981. *The Pans of Southwest Ecuador.* Paper presented at the meeting of the Society of American Archaeology. Pittsburgh.

———. 1982. "Aprovechamiento Agrícola Precolombino de Tierras Inundables en el Norte de Veracruz." *Biótica* 7(3): 343–357.

———. 1982. "Prehispanic Agricultural Use of the Wetlands of Northern Belize." In *Maya Subsistence.* Edited by Kent V. Flannery. New York: Academic Press.

———. 1983a. "Modelling Pre-Hispanic Hydroagriculture on Levee Backslopes in Northern Veracruz, Mexico." In *Drained Field Agriculture in Central and South America.* Edited by J. P. Darch. BAR International Series 189. Oxford: BAR.

———. 1983b. "Oriented Raised Fields in Central Veracruz." *American Antiquity* 48(1): 85–102.

———. 1989a. "Reducing the Risk: Some Indications Regarding Prehispanic Wetland Agricultural Intensification from Contemporary Use of a Wetland/Terra Firma Boundary Zone in Central Veracruz." In *Agroecology: Researching the Basis for Sustainable Agriculture.* Edited by Stephen R. Gliessman. New York: Springer.

———. 1989b. "The Persistence, Elaboration and Eventual Modification of Humboldt's Views of the Lowland Tropics." *Canadian Journal of Latin American and Caribbean Studies* 14(27): 85–101.

———. 1989c. *Tierra Configurada.* Mexico City: Consejo Nacional para la Cultura y las Artes.

———. 1990. *Between the Summit and the Sea.* Vancouver: University of British Columbia Press.

———. 1992. "A Favored Place: An Interpretation of the Development of a Wetland Landscape in Central Veracruz, Mexico." In *Person, Place* and Thing: Interpretive and Empirical Essays in Cultural Geography. Edited by Shue Tuck Wong. Geoscience and Man, Vol. 31. Baton Rouge: Department of Geography and Anthropology, Louisiana State University.

————. 1994. *El Mirador: A View from Above.* Paper presented at symposium, Organización Social y Representación del Espacio, Jalapa, Mexico, September 26–28.

————. 1995a. "Land-Use Succession in the Gulf Lowlands of Mexico." In *Global Land Use Change: A Perspective from the Columbian Encounter.* Edited by B. L. Turner et al. Madrid: Consejo Superior de Investigaciones Científicas.

————. 1995b. *Remontando el Río, de Nuevo.* Paper presented at symposium, V Encuentro: Los Investigadores de la Cultura Maya, Campeche, Mexico, Nov. 13–17.

———— and Maija Heimo. 1996. *Resistance to Agricultural Globalization.* Paper presented at Meeting of the Canadian Association for Mexican Studies, Mexico City, Nov. 10–13.

————, ed. 1977. *The Río Hondo Project. An Investigation of the Maya of Northern Belize.* Special Issue, *Journal of Belizean Affairs* 5.

Siemens, Alfred H., and Lutz Brinckmann. 1976. "El sur de Veracruz a finales del siglo XVIII—Un analysis de la 'Relación' de Corral." *Historia Mexicana* 26 (2): 263–324.

Siemens, Alfred H., and D. E. Puleston. 1972. "Ridged Fields and Associated Features in Southern Campeche: New Perspectives on the Lowland Maya." *American Antiquity* 37 (2): 228–239.

Siemens, Alfred H., Richard H. Hebda, Mario Navarrete Hernández, Dolores R. Piperno, Julie K. Stein, and Manuel G. Zolá Báez. 1988. "Evidence for a Cultivar and a Chronology from Patterned Wetlands in Central Veracruz, Mexico." *Science* 242: 105–107.

Skerritt Gardner, David. 1989a. *Una historia agraria en el centro de Veracruz.* Xalapa: Universidad Veracruzana, Centro de Investigaciones Históricas.

————. 1989b. "La 'modernidad' y el 'progreso' en el campo: el corredor central del estado de Veracruz en el siglo XIX. La Palabra y el Hombre." *Revista de la Universidad Veracruzana* 72: 111–136.

————. 1993a. "Colonización y modernización del campo en el Centro de Veracruz." *Siglo XIX: Cuadernos de Historia* II/5: 39–57.

————. 1993b. *Rancheros sobre tierra fértil.* Xalapa: Universidad Veracruzana.

Sluyter, Andrew. 1990. *Vestiges of Upland Fields in Central Veracruz: A New Perspective on Its Precolumbian Human Ecology.* Master's thesis, University of British Columbia. Available from National Library of Canada Microfilms, Ottawa.

————. 1994. "Intensive Wetland Agriculture in Mesoamerica: Space, Time and Form." *Annals of the Association of American Geographers* 84(14): 557–584.

————. 1995. *Changes in the Landscape: Natives, Spaniards, and the Eco-*

logical Restructuration of Central Veracruz, Mexico during the Sixteenth Century. Ph. D. Dissertation, University of Texas at Austin.

Sluyter, Andrew, and Alfred H. Siemens. 1992. "Vestiges of Prehispanic, Sloping-field Terraces on the Piedmont of Central Veracruz, Mexico." *Latin American Antiquity* 3(2): 148–160.

Soto E., Margarita, and Enriqueta García. 1989. *Atlas Climatologico del Estado de Veracruz.* Xalapa, Veracruz: Instituto de Ecología.

Stanish, Charles. 1994. "The Hydraulic Hypothesis Revisited: Lake Titicaca Basin Raised Fields in Theoretical Perspective." *Latin American Antiquity* 5(4): 312–332.

Stark, Barbara L. 1974. "Geography and Economic Specialization in the Lower Papaloapan, Veracruz, Mexico." *Ethnohistory* 21: 199–221.

———. 1978. "An Ethnohistoric Model for Native Economy and Settlement Patterns in Southern Veracruz, Mexico." In *Prehistoric Coastal Adaptations.* Edited by Barbara L. Stark and Barbara Voorhies. New York: Academic Press.

Stein, Julie. 1986. "Coring Archaeological Sites." *American Antiquity* 51(3): 507–527.

Stresser-Péan, Guy. 1968. "Ancient Sources on the Huasteca." In *Handbook of Middle American Indians,* Vol. 2, Part 2: *Archaeology of Northern Mesoamerica.* General Editor, Robert Wauchope. Austin: University of Texas Press.

Taylor, Bayard. 1850. *Eldorado, or the Adventures in the Path of Empire.* 2 vols. London: R. Bentley; New York: G. P. Putnam.

Thompson, J. Eric. 1970. *Maya History and Religion.* Norman: University of Oklahoma Press.

Thrower, Norman. 1990. "J. W. Humboldt's Mapping of New Spain (Mexico)." *Map Collector* 53: 30–35.

Tichy, Franz. 1979. "Genetishce Analyse eines Altsiedellandes im Hochland von Mexico. Das Becken von Puebla-Tlaxcala." In *Festschrift zum 42. Deutschen Geographentag.* Göttingen: Deutschen Geographentag.

Torres Lanzas, P. 1900. *Relación de los mapas, planos, etc. de México y Florida existentes en el Archivo General de Indias.* Seville.

Trens, Manuel B. 1947. *Historia de Veracruz,* Vol. 2. Jalapa-Enriquez, Mexico: n.p.

Troll, Carl. 1959. *Die Tropischen Gebirge.* Bonn: Dümmlers Verlag.

Turner, B. L., II. 1985. "Issues Related to Subsistence and Environment among the Ancient Maya." In *Prehistoric Lowland Maya Environment and Subsistence Economy.* Edited by Mary Pohl. Papers of the Peabody Museum of Archaeology and Ethnology, Harvard University, Vol. 77. Cambridge.

———. 1991. Review of *Ancient Maya Wetland Agriculture: Excavations on Albion Island, Northern Belize* (Mary Deland Pohl, editor). *American Antiquity* 56 (4): 736–737.

———. 1993. "Rethinking the 'New Orthodoxy': Interpreting Ancient Maya Agriculture and Environment." In *Culture, Form and Place: Essays in Cultural and Historical Geography.* Edited by Kent Mathewson.

Geoscience and Man, Vol. 32. Baton Rouge: Department of Geography and Anthropology, Louisiana State University.

Turner, B. L., II, and Peter D. Harrison . 1983. *Pulltrouser Swamp: Ancient Maya Habitat, Agriculture, and Settlement in Northern Belize*. Austin: University of Texas Press.

Turner, B. L., II, Antonio Gómez Sal, Fernando González Bernáldez, and Francesco di Castri. 1995. *Global Land Use Change: A Perspective from the Columbian Encounter*. Madrid: Consejo Superior de Investigaciones Científicas.

Vadillo López, Claudio. 1994. *La Región del Palo de Tinte: El Partido del Carmen, Campeche 1821–1857*. Campeche, Mexico: Fondo Estatal para la Cultura y las Artes.

Vargas Pacheco, Ernesto. 1994. "Síntesis de la Historia Prehispánica de los Mayas Chontales de Tabasco-Campeche." *América Indígena* 1(2): 15–61.

———. 1995. *Proyecto Arqueológico "El Tigre."* Mexico City: Universidad Nacional Autónoma de México.

———. 1996. Personal communication.

Vázquez Yáñez, Carlos. 1971. "La vegetación de la laguna de Mandinga, Veracruz." *Anales del Instituto de Biología, Serie Botánica* 42(1): 49–94.

Venegas Cardoso, Francisco Raul. 1978. *Las Chinampas de Mixquic*. Thesis, Department of Biology. Mexico City: Universidad Nacional Autónoma de México.

Waibel, Leo. 1950. "European Colonization in Southern Brazil." *Geographical Review* 40: 529–547.

Walter, W. Grey. 1951. "Activity Patterns in the Human Brain." In *Aspects of Form*. Edited by Lancelot Law Whyte. Lund: Humphries.

Welcomme, Robin L. 1979. *Fisheries Ecology of Floodplain Rivers*. New York: Longman.

West, Robert C. 1970. "Population Densities and Agricultural Practices in Pre-Columbian Mexico with Special Emphasis on Semi-terracing." *International Congress of Americanists, Stuttgart-Munchen, 1968: Proceedings* 2: 361–369. Munich.

———. 1972. "The Relaciones Geográficas of Mexico and Central America, 1780–1792." In *Handbook of Middle American Indians*, Vol. 12, Part 1: *Guide to Ethnohistorical Sources*. Edited by Howard F. Cline. Austin: University of Texas Press.

West, Robert C., N. P. Psuty, and B. G. Thom. 1969. *The Tabasco Lowlands of Southeastern Mexico*. Baton Rouge: Louisiana State University Press.

Wheelock, Arthur K. 1989. *Still Lifes of the Golden Age*. Washington, D.C.: National Gallery of Art.

Wild, Alan, ed. 1988. *Russell's Soil Conditions and Plant Growth*. 11th ed. New York: Longman/Wiley.

Wilford, John Noble. 1980. "Maya Canals Found by Radar." *New York Times*, June 3, p. C3.

Wilk, R. R. 1981. *Agriculture, Ecology, and Domestic Organization among the Kekchi Maya*. Ph.D. dissertation, University of Arizona.

Wilkerson, S. J .K. 1980. "Man's Eighty Centuries in Veracruz." *National Geographic* 158(2): 203–231.

Williams, Edwin B. 1955. *Spanish and English Dictionary*. New York: Holt.

Wilson, Robert Anderson. 1855. *Mexico and Its Religion*. New York: Harper.

Wise, George S. 1952. *El México de Alemán*. Mexico City: Editorial Atlante, S.A.

Wright, A. C. S., D. H. Rommer, R. H. Arbuckle, and V. E. Vial. 1959. *Land in British Honduras*. Colonial Research Publications 24. London.

Wright, Ronald. 1989. *Time among the Maya: Travels in Belize, Guatemala, and Mexico*. New York: Viking.

Wright, William, Jr., and Frank J. Little, Jr. 1972. *Ecology and Pollution*. Philadelphia: North American Publishing Co.

Index

acahual, 104, 109
Acalán (Campeche), 103
Acrocomia mexicana, 53. *See also* palms
adaptation, 261
agrarian: code, 221; reform, 204, 220, 223, 259
agricultura aventurera, 224–225
agricultura básica, 196, 201, 232–233, 260. *See also* subsistence; "traditional" agriculture
agricultural: calendar, 231 tbl.; chemicals, 229; extension, 235–239, 237 tbl.; land, capability of, 221–228; products, 125 (*see also* crops); research, 212, 229, 235
agriculture: 125, 185–191, 212, 228 tbl., 230 tbl., 261; annual cycles of, 98–100; in 18th-century maps, 157; flood-recessional, 42; fugitive, 41; intensification of, 178, 257; interdigitation of, 55, 105, 106, 195; irrigated, 228 tbl.; *marceño*, 42; microenvironmental complementarity in, 98–100; "modernized," 214; on the piedmont, 177–179; Prehispanic (*see* platforms and canals; "traditional" agriculture); in the wet season, 98, 100
aguada, 224
Alemán, Miguel: 198, 199, 204; his attitude toward the Green

Revolution, 214; estate of, 205–207, 206 fig.
alluvial: lands, 224; plain, 50
alluvium: in *barrancas*, 225
altitudinal: differentiation of environments, 215, 251–252; land use, 190
altitudinal zonation: 168, 268; of Central Veracruz, 167–169, 171 fig.; of Humboldt, 152
animals, domestic: on ejidos, 234; goats, 143, 194, 242; health of, 239, 246; horses, 142, 242; mules, 194; numbers of, 241 tbl.; pigs, 242; poultry, 234, 242; on private properties, 242; sheep, 142, 143, 194, 242; in 16th century, 141, 142–148. *See also* cattle
animals, wild, 67, 121–123, 174
Apuntes para la descripción de Veracruz, 111
arcabuco, 123, 124 fig., 127 fig., 173
archaeological interpretation, 84
arroyos, 58
autoconsumo: extent of in 20th century, 232

bajo, 9 fig., 22–31
banana, 187–188, 229
barranca: 156, 178 figs. 6.6, 6.7, 179, 195, 257; cultivation in, 190, 224; Indians in, 181, 182;

pastures in, 179; profile of, 180
fig.; as refuge, 181; soils in, 223
Belize, Northern, 9 fig., 20–21, 264
bias. *See* representation
Blacks, 182
bosque tropical perennifolio, 124.
  See also selva
bottomlands, 226–228. *See also*
  wetlands

cacao, 186
camellones, 35
Campeche, Southwestern: 11–20;
  agricultural remains in, 12–15,
  28. *See also* El Tigre
canals: 16, 85; as fisheries, 18, 19
  fig., 67; midlines of, 74 fig., 76
  fig., 79 fig.; morphology of, 70,
  72, 77, 83–87; orientation of, 77,
  88–91; purpose of, 16–19, 36–
  38, 44, 97; transportation in, 16,
  17 fig., 29; water control in, 44
canícula, 56
carrying capacity, 86, 245
cartography, 114, 129–133, 155,
  200
cattle: 121, 144, 158, 159, 177,
  193–195, 246, 271; breeds of,
  242–243, 244, 246; on ejidos,
  234–235; numbers of, 239 tbl.,
  240–242; perceptions of, 211;
  water for, 42 fig. *See also cima-*
  rrones
Cedrela mexicana, 124
Cempoala, 95, 103
Central Mexico, 171 fig.
Central Veracruz: 169–170; altitudinal zonation of, 167–169, 171
  fig.; maps of, 153 fig., 154 fig.;
  physiography of, 50–60
ceramics, in agricultural vestiges,
  90, 92, 93, 227
chaparral, 123, 172 fig., 173
charcoal, 94, 234, 235
Charles III, 149
chile, 229
chinampa cultivation: 253; in

Campeche, 14; in the lowlands,
  44; and soil fertility, 68; technique of, 10, 42–43; and water
  control, 87
chronology: 95–96; of Central
  Veracruz, 101–102; of wetland
  agriculture, 90, 92–94, 106, 253,
  255
cimarrones, 111, 143, 145–147,
  158–159, 194, 259
climate: 108; in Central Veracruz,
  53–57; change of, 43; in 16th-
  century documents, 115–119;
  and winds, 54, 55, 171
coffee, 187
Comisión Nacional de Irrigación,
  197
Companía Mexicana de Aerofoto,
  200
comparative advantage, 249
complementarity: of environments, 77, 83, 247, 252; of production, 83; of sources of investigation, 264. *See also*
  interdigitation
complementary income, 254
Conquest, impacts of, 107–109
continuity of sustaining activities,
  261–262
Convolvulus batata, 229
co-operatives, 212–213
cosmographic interpretation, 267–
  268
cosmography: 255; of canals, 88–
  91
cotton, 128, 129 fig., 187
credit for agriculture, 212
crops: 187; annual calendar of, 231
  tbl.; in the dry season, 229, 230;
  extent of, 230 tbl. *See also under*
  individual crops
Cyperus s., 53 fig.

dairy farming, 226. *See also* milk
dams: in Candelaria river, 15; as
  causeways, 14; in wetland fields,
  87

de Alvarado, Pedro, 105
Declaración sobre Veracruz, 111,
112, 128, 261
del Corral, Miguel: 148, 149,
150, 157–158, 265; map of
Central Veracruz by, 153 fig.;
map of Southern Veracruz by,
156
Denevan, William, 263
depopulation, 107–109, 133, 134–
135, 256. *See also* Indians
development, 203–204, 205, 261,
270. *See also* modernization
Díaz del Castillo, Bernal, 102–104,
107
disease: 161, 218; causing depopulation, 108. *See also* malaria; yellow fever
disintensification. *See*
intensification
documents. *See under individual*
documents
drainage: as goal, 203; modern, 97;
need of, 213; of wetlands, 226,
248
drought, 57
dry season, 234
dunes: 100, 171; soils of, 225; vegetation on, 172–173, 210

Echinocloa sp., 53 fig.
economy: in 19th century, 188–
189; on small farms, 232–233; in
20th century, 212–213
ecotone: 9, 38, 39; and inundation,
14; and ranching, 147
18th-century documents: 148–155,
157 fig.; on agriculture, 157; on
hydrography, 156; maps in, 152
fig., 154 fig.; on ranching, 158–
159; on vegetation, 148, 149,
156–157
ejido: 220, 259; and agrarian reform, 204; and animals, 241 tbl.;
and land tenure, 38, 196, 213,
242; size of, 220; of Vargas, 233–
235

El Palmar, 40, 42 fig., 254
El Tajín, 95
El Tigre: agricultural remains at,
12–15, 28; chronology of, 12
entrepreneurship, 203, 204
environmental: change, 64, 254;
zonation, 215
epoca negra, 189
estancia, 102, 128, 129 fig., 131,
141, 142, 143–144, 226, 256,
257
ethnography, 5
excavations, 91–95

farm size, 237–238 tbl.
fertilizers, 211, 230
Ficus, 53
figs, 53
fish: 120–121; habitats of, 67, 200
fishing: 18 fig., 61, 62, 66, 67–68,
87, 99, 254; as part of subsistence, 66 fig.
flooding: 61–70; affected by modern canals, 199, 200; depicted
in murals, 70; duration of, 59,
62, 65; effect on cattle, 247–
248; effect on fish, 65; extent
of, 61–64, 65; as fallow, 67–69;
phases of, 65–70. *See also*
inundation
floodplain: 40–43; fish in, 61–62,
67; seasonal land use on, 66 fig.
foreign observers, 5–6, 158
forest: and grazing, 243, 245; clearance of, 234, 243; extent of, 228
tbl.; gallery, 127 fig., 173; *monte*
243; tropical, 38, 39. *See also*
selva
form of vestiges, 83–87, 255
fruits: 125; extent of, 230 tbl. *See*
also under individual plants;
orchards
fruit trees: 42 fig., 229; extent of,
230 tbl.
function of canals: microenvironmental, 9; vs. form, 87; vs. terminology, 87. *See also* models

gardens: 99, 187; in *aguadas*, 224; on ejidos, 234. *See also* kitchen gardens

geology, 222 fig.

globalization: adverse effects of, 8, 249, 263; and NAFTA, 203; resistance to, 252

de Gómara, López, 105, 107

grasses: 230 tbl.; varieties of, 245; *zacat*, 173

grasslands, 126–128

grazing: 110–111, 243, 245; on croplands, 245; effect of, on grasslands, 126; seasonality of, 66 fig.

Green Revolution, 214

Gynerium saccharoides, 173

hacendado, 240

hacienda: 155, 259; Acazónica, 192–193, 205; *del ganado*, 192; El Faisán, 196, 223; El Mirador, 163, 165, 166, 186, 196, 269; Manga de Clavo, 175, 205; as *panopticon*, 165

harvest, timing of, 100

hazards: 57; natural, 252

Hernández, Arias, 110, 111

Hernández Diosdado, Alonso, 111–112

hill lands: agriculture on, 179, 181; dry season activities on, 100; interdigitation of, 235; vegetation on, 210

Huasteca: 35–35, 106; and wetland agriculture, 95

hunting: 121–122, 190, 254, 271; of feral cattle, 146; resources of, 234; in various environments, 67, 99, 177

hurricanes, 57, 252

hydraulic intervention: 197–200; context of, 203–207; cost of, 210

hydrography: of Central Veracruz, 57–60; in 18th-century maps, 156

hydrology: changes in, 43; control of, 10–14; in 16th-century documents, 119–120

hydrophytes, 42 fig., 53 fig., 127 fig.

hydrosere, 44

Hymenachne amplexicaulis, 53 fig.

income: 218; on ejidos, 233, 234

Indian communities, 133–135, 155

Indians: adaptation of, 181, 182; in *barrancas*, 182; in colonial period, 256; depopulation of, 133, 134–135, 256; in 19th century, 181–182; representation of, 134, 182, 183 fig., 256; settlements of, 133–135; subsistence of, 182; and vanilla cultivation, 186

ingenieros: 201–202, 207–208, 266; and their perceptions, 207, 210, 215, 248, 270

integration: of agricultural activities, 233, 261; of agriculture and wage labor, 233; of cropping and animals, 214

intensification: of agriculture, 100, 178, 257; disintensification, 12, 44, 253–254; on floodplains, 66 fig.; on the piedmont, 83; in Prehispanic times, 266; process of, 10, 14, 15; as progress, 210; of ranching, 254–255; of water use, 185; of wetland agriculture, 44, 94, 253–254

interdigitation: of environments, 177, 231, 271; of microenvironments, 252; of productive activities, 261; in the savanna, 177. *See also* complementarity

intermediaries, 212, 232, 235

inundation: margins of, 14, 82; on river terraces, 224; range of, 18; role of, in ranching, 249; timing of, 97. *See also* flooding

invasion of land, 240

irrigation: 42 fig., 203–205, 237 tbl., 239; canals for, 197–200; effect of, on land tenure, 221;

effect of, on soil quality, 227; efficiency of, 210; extent of, 228–229 tbl.; by gravity, 215; of pastures, 255; potential for, 249; from Prehispanic canals, 87; soil capability for, 225; on wetland margins, 248

irrigation canals: as boundary of a region, 25; modern, 209; in the San Juan Basin, 197–200

irrigation district: Actopán, 201, 209, 215; La Antigua, 201, 209, 210, 215, 216 fig., 217 fig., 228, 230 tbl.; numbers of domestic animals in, 241 tbl.

irrigation water: contamination of, 200; use of, 210

Itzamkanac. *See* El Tigre

Jalapa: 265; climate of, 55 fig.
Jarcoho, 184

karst: 13 fig.; hydrology of, 18–20, 24–26; polje, 24–26

kitchen gardens: 109, 125, 211. *See also* gardens

knowledge, "traditional," 234

La Antigua, town of, 61, 103, 129, 173. *See also* rivers

labor: in ranching, 239–240; types of labor force, 218, 239–240, 245; in wetland agriculture, 85–86

Laguna Catarina: 45, 48, 49 fig., 59, 60 fig., 62; chronology of, 93; in 18th-century maps, 156; extent of, 222

lake margins, 42–43

land, 240

land capability, 221–228. *See also* soil classification

land ownership: 38, 39; altitudinal, 247; and legislation, 204, 249

landscape degradation after Spanish contact, 109

land tenure: 204, 205, 213, 220–

221, 237 tbl., 238; and Prehispanic fields, 257–259

land use: 34, 271; altitudinal, 190; interdigitation of, 66 fig.; irrigated, 225–226, 228 tbl.; models of, 38–44; seasonality in, 96; and soil series, 226; in various epochs, 256–257, 259, 271; in wetlands, 42, 83

Leersia hexandra, 53 fig.

Leguminosae, 173

levées: 105, 109; land use on, 38–39; settlements on, 36, 37 fig.

liberalism, 164

logging, 157

lowlanders: 181; naming of, 145; representations of, 152, 161, 179, 214, 216

lowlands: agriculture on, in 19th century, 185–191; at Contact, 101–109; ownership of, 240; representations of, 157–158, 271; sedimentation in, 171, 172 fig.

maguey, 100, 128

maize: 42 fig., 124; annual calendar of, 231 tbl.; cultivation of, 230 tbl., 233–234; in 19th century, 187; pollen of, 92; staggered production of, 211

malaria, 99, 117–119. *See also* disease; yellow fever

Mammea americana, 124–125

mango: 229; and land capability, 223

maps: of Central Veracruz, 130 fig., 132 fig.; as paintings, 114; in 16th-century documents, 114–115, 129–133

market production, 232

marl, 92

matorrales, 52

Maya: culture regions, 12; groups, 11, 20; lowlands, 8–9, 11–12, 27; subsistence base of, 30–31

Medellín, town of, 124, 135, 265

Medellín Zenil, Alfonso, 266

medical plants, 134
microenvironments: complementary use of, 195, 247; differentiation of, 252; interdigitation of, 15, 252, 259
mid-lines of canals: 70, 72, 74 fig., 76 fig., 79 fig., 85, 88, 255; as indicators of morphology, 83, 86. *See also* canals
milk, 145, 242, 246
milking stations, 127 fig.
milpa, 42 fig.
minifundia: 213, 220; and *autoconsumo*, 232
models: of altitudinal zonation, 252; of canal function, 36–38, 38 fig.; of intensification, 44; of location of patterning, 82–83; of ranching, 144; of wetland use, 36–44
modernization: 236; adverse effect of, 8, 214, 249; effect on subsistence, 67; perceptions of, 203, 210–212, 236; program of, 213–215; of ranching, 249; in various epochs, 196–200, 202–203
morphology. *See* form
multiple cropping, 229
municipalities, 215–216

NAFTA, 203
New York Times, 29
19th-century documents, 257, 259, 269
nortes, 55, 57, 100, 140, 229, 252

Olmec: heartland at Spanish contact, 104–105; relationship to wetland cultivation, 95–96
orchards, 228 tbl.
orchestration of microenvironments, 252. *See also* complementarity; interdigitation
orientation: agricultural significance of, 90; of canals, 72, 77, 88–91, 255; and cultural

landscape, 91; of stone lines, 82, 90, 100
Orive Alba, Adolfo, 199, 202

Palerm, Angel, 9, 11
palms: 72, 173–174, 175; coco, 230 tbl.; *Cocos de Guinea*, 123, 124 fig.; in colonial documents, 53; *coyol*, 53, 124; *Roystonea dunlapiana*, 53 fig., 69; in 16th century, 123–124
panopticon, 165
pasture: 42 fig., 228 tbl.; in canyons, 129 fig., 175; in wetlands, 175
patrón-client relationships, 235
patterning: 255; as an artifact, 85; context of, 37–38; distribution of, 49 fig.; and hydrology, 82–83; location of, 8–9; after Spanish contact, 109. *See also* vestiges
pedology, 221–228
Petén, 22, 24, 25 fig.
physiography of Central Veracruz, 50 fig.
phytoliths, 91, 92
piedmont: 80, 81 fig., 82 fig.; agriculture on, 82, 83; dry season activities on, 100; vegetation on, 128, 178 fig.; water management on, 177–179
pinturas, 114–115, 151–152
place names, Indian, 155
plants, taxonomy of, 268. *See also* crops; fruits; wood trees
platforms and canals: 10, 43; abandonment of, 67; construction of, 255; extent of, 86, 99; function of, 24, 255; maintenance of, 99; in 19th century, 175; post-abandonment changes of, 85; workload, 85. *See also* Prehispanic wetland fields; proto-chinampa; raised fields
polje, 24
Pontederia sagitatta, 54 fig. *See also* hydrophytes

population, 216, 217, 271
Post-Classic in Central Veracruz,
    101–102
precipitation, 53–57, 56 fig.
Prehispanic settlements: chronol-
    ogy of, 101–102; and soil series,
    224; water storage in, 224
Prehispanic wetland fields: and
    modern canals, 199; vs. modern
    land tenure, 257, 258 fig., 259.
    See also platforms and canals;
    raised fields; wetland agriculture
presidents: Miguel Alemán, 198,
    199; Santa Ana, 175, 193
private properties, 220, 241, 242
productivity: of lowlands, 271–
    272; of wetlands, 270–271
proto-chinampa, 10, 43, 253
Pulltrouser Swamp, 21, 24

Quintana Roo, 22–23

radar, 27–31
rainfed agriculture, 228, 238 tbl.
    See also agriculture
raised fields: 252–256; and in-
    tensification, 252; role of, in Pre-
    hispanic agriculture, 252–253;
    role of, in subsistence, 15; termi-
    nology of, 30, 43. *See also* plat-
    forms and canals
ranchería, 155, 184–185
ranchero: 239–240; occupations of,
    184–185; representation of,
    184–185, 191
ranching: and dairy farming, 226;
    and environmental complemen-
    tarity, 254–255; extension for,
    235–239; family's role in, 239–
    240; intensification of, 254–255;
    modernization of, 249; percep-
    tions of, 211, 246; practice of,
    245–246; on private properties,
    235–248; seasonality of, 96; and
    topography, 226; in various
    epochs, 127, 141–147, 158–159,
    191–195, 211, 229, 245–246

rancho: 196, 259; location of, 127
    fig.; in 19th century, 192–193
refuge, 181
regionalization of the San Juan
    Basin, 215–216
*Relación de la Ciudad de Veracruz
    y su Comarca*, 111, 112, 114–
    115, 128
Relación of 1744, 151
Relación of 1771: 149, 150; repre-
    sentation of lowlands in, 157–
    158
Relación of 1777: 148, 150; *pin-
    turas* in, 151–152; representa-
    tion of lowlands in, 148, 158
Relaciones Geográficas, of 16th
    century: 111, 150, 261; nature in,
    115–128; purpose of, 112; as re-
    search material, 269, 270
religion, 136, 219, 240
representation: 4–5, 6, 146, 184,
    215, 263–272; of colonial land-
    scapes, 131; of lowlanders, 161,
    179, 181, 188, 214, 216; of low-
    lands, 114–118, 207, 213, 215;
    by maps, 114; misrepresentation,
    28, 31; of races, 134, 182, 183
    fig.; of rancheros, 184–185, 191;
    of religion, 219; romantic in-
    fluences in, 269; of the San Juan
    Basin, 215; in Spanish accounts,
    106; of tropics, 161; of wetland
    agriculture, 209–213; of wet-
    lands, 175, 196, 209–210, 248
Richards, Paul, 261
ridged fields, 43. *See also* platforms
    and canals
risk, 42, 235
river-basins, development of, 203–
    204
river margins, 224
rivers: 9 fig., 37 fig.; Candelaria,
    11–20, 68; Coatzacoalcos, 32–
    33, 105; Estero de Tres Bocas, 36;
    Grijalva, 18, 32; Hondo, 68, 20–
    21; La Antigua, 45, 57, 58 fig.,
    59, 60 fig., 61, 63 fig., 71 fig.,

119–120, 124 fig., 130 fig., 171,
178, 197, 209, 223, 224; Medellín, 130 fig.; Nautla, 35, 36; New,
20–21, 45, 57, 61; Papaloapan,
33, 105–106; Paso de Ovejas, 58,
59, 61, 174, 223, 224; San Francisco, 58; San Juan (Central Veracruz), 46 fig., 58, 124 fig., 127
fig., 209, 264; San Juan (Southern
Veracruz), 33; Tamiahua, 35–36;
Tecolutla, 35; Túxpan, 35, 36;
Usumacinta, 18, 32, 104; water-
levels in, 33
river terraces, 224
romanticism, 170, 171, 181, 269
Roystonea dunlapiana, 53 fig., 69
Rugendas, Moritz, 169–170, 176
fig.

salinity of soil, 227
Salix, 53
San Juan Basin: chronology of wetland agriculture in, 92–94; flooding in, 46 fig., 63 fig., 199, 200;
hydraulic works in, 198 fig.; hydrography of, 57–60, 60 fig.; Laguna Catarina in, 45; land tenure
in, 205; land use in, 228 tbl.;
physiography of, 48–60; as a region, 215; in 20th-century documents, 200–202; vertical zonation in, 48
San Juan de Ulúa, 136, 139–141,
265
Santa Ana, 175, 193, 205
Sartorius, Carl: 257, 266, 269;
analyses of, 162–164; background of, 163, 165; biases of,
164–165; on colonial rule, 188;
on commercial crops, 185–188;
hacienda of, 163, 165; as a mod-
ernizer, 196; publications by,
163, 167
savanna: 123, 124 fig., 127, 127 fig.,
129 fig., 177–178, 257; origin of,
126; in 16th-century documents,

126; and soil series, 226; vegetation of, 178 fig.
Sayula: estate of, 205–206, 223;
town of, 206
Scheelia liebmannii, 53. See also
palms
Scholes, France V., 103
science, 6, 267, 268
scooping, 87. *See also* irrigation
sea level, 64
seasonality: 230, 245; effect of on
ranching, 235, 247; on a floodplain, 66 fig.
seasons: agricultural, 98–100; in
the savanna, 177
*Secretaría de Agricultura y Recur-
sos Hidráulicos,* 200
*Secretaría de Recursos Hidráuli-
cos:* 45, 46 fig., 60 fig., 198 fig.,
200, 207; head of, 199; projects
of, 196–200, 204
sedimentation: on floodplains, 68;
in lowlands, 172 fig.; in wetlands, 43
sediments: effect of on floodplains,
68; in flood waters, 65, 68; transport with flooding, 68
seepage, 20, 100
selva (baja caducifolia): 42 fig., 52,
53 fig., 83, 123, 126, 128, 129
fig.; *mediana perennifolia,* 83;
mediana subcaducifolia, 72;
mediana subperennifolia, 39,
52, 53 fig. See also *bosque*
settlements: 217 fig.; estancias,
143–144; of Indians, 133–135;
on levées, 36, 37 fig.; location of,
in regard to patterning, 77, 95–
96; on piedmont, 177–179; Santa
Elena as, 36; in the savanna, 177;
in 16th century, 128–141, 143–
147; in the wetlands, 174
shifting cultivation: 10, 109, 187,
229; representation of, 210; role
of in Prehispanic subsistence,
9–10

Sístema Alimentario Mexicano, 97
site of investigation: 9 fig.; *bajos*
22–23; Candelaria, 12–15, 68;
Caño Prieto, 77, 79, 80, 80 fig.;
Carmelita, 72, 74, 75 fig., 76 fig.;
El Palmar, 97; El Yagual, 2, 6 fig.,
77, 90, 93 tbl.; locational map,
17 fig.; Nevería, 64, 77, 78 fig.,
79 fig., 97; New and Hondo
rivers, 20–21; Petén, 25 fig.;
Pulltrouser Swamp, 21; Rinconada, 80, 81 fig., 82; Tulipan,
72, 73 fig., 74 fig., 90, 97
16th-century documents: 102–109,
110–114, 128–141; maps in,
114–115; on nature, 115–128;
on ranching, 141–147
slavery: 107, 182; in ranching, 144,
145; in Veracruz, 136
soil classification: *fluvisoles*, 222;
micro-altitudinal, 270; in the
San Juan Basin, 209, 221–228;
tierra de humedad, 222, 223
soil moisture and flooding, 67
soils: drainage of, 224; irrigation
capability of, 222; and land use,
226; water storage capability of,
227
soil series: 222–228, 222 fig.
Sorgo escobero, 234
springs, 14, 20, 42
starting bed, 42 fig.
stone lines: 81 fig., 82 fig., 178; on
field boundaries, 80; orientation
of, 82, 90, 100
stratigraphic sequence, 5, 92
subsistence: 232–233; characterization of, 259–260; continuity
of, 261–262; definition of, 232;
of the Maya, 30–31; perception
of, 232; in *Relaciones Geográficas*, 116; and staggered production, 210; in various epochs,
189–191, 232
succession, of landscapes, 147
sugarcane: 124 fig., 128, 185, 186,

229, 259; in canyons, 129 fig.; extent of, 230 tbl.; processing of,
184; and soil series, 223
sugar industry, 218
sustainability, of traditional agriculture, 263

Tabasco, 104, 186
taxonomy, of plants, 268
tectonic activity, 43, 64
temporal, 233
tenezquauitl, 124
tenure, of wetland agriculture, 255.
See also land ownership; land
tenure
Teotihuacán, 89–91, 96
terminology, of wetland agriculture, 43, 87
terracing, 10
terra firma: grasslands on, 126–
128; and intensification in the
wetland, 44; land-use on, 38,
66 fig.
text, as a metaphor, 4–5, 84, 263–
264
tierra de humedad, 222, 223, 224
Titicaca, Lake, 8
tobacco, 186
tomato, 229
tonalmil, 97
topography, to the west of Veracruz, 47 fig.
Totonaca, 95
trade: 98; in 19th century, 185–
187, 188, 189; in 16th century,
138–139
"traditional" agriculture: 211;
characterization of, 260–261; diversity in, 231 future of, 249; potential for development, 261; in
various epochs, 260
transhumance: 96, 110, 142–143,
144, 247–248; arrangement for,
235; movements in, 195
transnationalization, 249
transportation: along canals, 16,

99; routes in 16th century, 136; seasonality of, 66 fig.
travel accounts: 5–6; deconstruction of, 6; in 19th century, 161–162; in 16th century, 113–114
travelers: 5–6, 160–161; German, 161–162; in 19th century, 45
tribute, 97
tropics, representation of, 161
Tuxtlas (region), 55, 93, 108
20th-century documents, 257, 259, 270

vanilla, 186
vaquero, 185, 193, 194
vegetables: annual calendar of, 231 tbl.; extent of cultivation, 230 tbl. *See also under individual plants*
vegetation: 38, 39, 52, 53, 83; aquatic, 53; *arcabusco*, 173; in Central Veracruz, 52–53; *chaparral*, 172 fig., 173; on the dunes, 52, 172–173; in 18th-century maps, 156–157; hydrophytes, 175; palms, 53, 173–174, 175; on the piedmont, 128; in the *Relaciones*, 148, 149, 157; savanna, 177–178; in 16th century, 123–128, 124 fig., 127 fig., 129 fig.; at Spanish contact, 109; various species, 173; in the wetland, 52, 98, 173–176; on the wetland margin, 42 fig. *See also* palms; *selva*; wood trees; *and under individual species*
Veracruz, city of: location of, 135, 136, 140; old Veracruz, 265; as a sick place, 135; in 16th century, 135–139
Veracruz, Northern, 35–36
Veracruz, port of, 182
verticality, 251–252, 257
vertisols, 22, 26
vestiges, relationship of, to topography and hydrology, 14. *See also* patterning

volcanism, 179
von Humboldt, Alexander: 161, 269, 270–271; altitudinal scheme of, 152; map of Central Veracruz by, 152, 154 fig., 155; profile of Central Mexico by, 156, 157 fig.; as scientific authority, 162, 167; views of, regarding the lowlands, 152, 203

wage labor, 233, 254
water control: catchment depressions for, 42; extent of, 253; along wetland slope, 44. *See also* hydraulic intervention; hydrology
watering, of cattle, 245–246, 247
water levels: fluctuations of, 38, 43, 58 fig., 65; in the San Juan Basin, 59
water management in wetland agriculture, 255; on the piedmont, 177–179; Prehispanic, 185
water storage, 14, 77
wells, 97–98
wetland agriculture: abandonment of, 102, 106; adjustments in, 64; in Andean South America, 11; chronology of, 21, 24, 90, 94, 106, 253, 255; construction of, 255; disintensification of, 44, 253–254; evolution of, 10, 44; expansion of, 86; forms of, 253, function of, 253, 255; intensification of, 44, 253–254; and microenvironmental complementarity, 254–255; models of, 82–83; morphology of, 255; orientation of, 255; policy implications of, 8; and other production, 254–255; reactivation of, 7–8; role of, in subsistence, 15, 256; and soil fertility, 68; tenure of, 255; terminology of, 43, 253; water management in, 95–96, 253, 255. *See also* chinampa; platforms and canals;

Prehispanic wetland fields; water management
wetland boundary: 13 fig., 109, 248; favor of, 205; land use on, 42 fig., 97–98; model of, 38 fig., 41 fig.
wetlands: agriculture in, 179, 181; attitudes toward, 6–7, 203, 264–267; characterization of ecology, 174–175; clearance of, 127; dry season activities in, 99; favor of, 205; productivity of, 270–271; ranching in, 127, 248; sedimen-tation in, 43; settlements in, 174; as source of illness, 6; vege-tation in, 173–176, 176 fig.; and watering of cattle, 247
wildlife, 120–123, 174, 177. *See also* animals, wild
winds, 229. *See also nortes*
wood trees, 53, 173, 124–125

yagua, 53 fig., 69
yellow fever, 99, 116, 166, 170. *See also* disease